PRACTICING CASTE

FORDHAM UNIVERSITY PRESS NEW YORK 2019

COMMONALITIES
Timothy C. Campbell, series editor

PRACTICING CASTE

On Touching and Not Touching

ANIKET JAAWARE

Fordham University Press has no responsibility for the persistence or accuracy of URLs for external or third-party Internet websites referred to in this publication and does not guarantee that any content on such websites is, or will remain, accurate or appropriate.

Fordham University Press also publishes its books in a variety of electronic formats. Some content that appears in print may not be available in electronic books.

Visit us online at www.fordhampress.com.

Library of Congress Cataloging-in-Publication Data available online at https://catalog.loc.gov.

Printed in the United States of America

21 20 19   5 4 3 2 1

First edition

CONTENTS

Foreword by Anupama Rao vii

Introduction 1

1. Touch and Its Elements and Kinds 11

2. Touch—An A Priori Approach 37

3. Touch in Its Social and
   Historical Aspects I 61

4. Touch in Its Social and
   Historical Aspects II 93

5. Touch and Texts: Ancient and
   Modern 119

6. (Un)touchability of Things and
   People 148

7. Society, Sociality, Sociability 170

8. Recapitulation with Variations 190

   Coda 205

   Notes 209

   Bibliography 223

   Index 233

FOREWORD

Such a promiscuous word *caste* is. It is used in so many ways and brings so many references, and it delivers, after some yielding and coaxing, so many meanings. It does not remain faithful/pure/chaste to itself ever; it is clear that no caste is celibate.

So begins the last chapter of *Practicing Caste: On Touching and Not Touching*. The book is a remarkable exercise in showing what is possible when we attend to caste as if we were confronting it for the first time, bracketing the extensive scholarship about it, and refusing to adopt a ready political stance against caste discrimination and inequality. *Practicing Caste* asks what new ways of thinking about caste are enabled when we approach "caste" ignorantly, that is, when we forget the weight of its millennial history and turn to caste less as an exception than an occasion to rethink the grounds of sociability.

The social, sociality, and sociability are very much at issue in this work, which is marked by acute ethical purpose. However, Aniket Jaaware begins by setting aside questions of ethics and politics. Instead, he invites us to undertake an experiment: Make caste less abstract and less *Indian* by letting go of the standard paradigms—literary, sociological, philosophical, Indological—that have allowed caste to function as an analytic shorthand for understanding a social order, a set of practices, and a way of thinking. To defamiliarize caste and mark it as one form of social distinction among others brings caste within the fold of (global) philosophical inquiry as one iteration of the more general problem of "the human." *Practicing Caste* also challenges those who argue from the perspective of caste's apparent facticity to ground their analysis in first principles.

Oublierring (*oublier* + *err*) is the term Jaaware has coined to describe this act of forgetting and thinking anew. He urges us not to restrict our understanding of caste to the performance of hierarchy as anthropologists do, engage with it as a form of stratification and inequality as sociologists insist, or to approach caste as a form of community and identity as many historians argue.[1] Instead, caste is removed from the domain of subcontinental specificity and associated with the simple divide between touching and not touching, of bringing near or keeping something or someone at a distance. The text brilliantly argues that the peculiar (and peculiarly contingent) manner by which caste became "Indian," and the subcontinent imagined as a place governed by the enduring consequences of ritualized forms of inclusion and exclusion, can be traced back to the simple distinction between touching and not touching, rather than any originary ban, taboo, or proscription. Caste is concrete; it is generated by touch: touching yourself, fondling a baby, the chair touching your buttocks, killing someone. Touch is everywhere, and touch is "happening" all the time whether willingly or unwillingly. The ubiquity of touch allows it to withdraw from consciousness, even as it grounds perception.

How are forms of agonistic intimacy generated by the archaic and somewhat anarchic facticity of touch? How do they take form and shape as that (impossible) intimacy we apprehend as "caste"?

Perhaps we can best appreciate the question, and the distinctive nature of Jaaware's response to it, by bringing *Practicing Caste* into conversation with the work of the radical anticaste thinker B. R. Ambedkar (1891–1956), who struggled his entire life to comprehend the astonishing longevity of a social order governed by proscriptions on touch, which was manifest in the absolute divide between those categorized as "Touchable" and "Untouchable." *Practicing Caste* questions the relationships among caste, touch, and untouchability, but it does so by avoiding sociologism and leaving open-ended the question of how identity relates to experience.

Ambedkar's effort to understand the system of caste focused on the social regulation of sexual touch, which translates into a proscription on marriage across castes. (In kinship studies, this is the concept of caste endogamy.) Ambedkar asks: If "caste" is predicated on the preordained legibility that marks the child by the caste of his or her father, then what is the relationship between "sex" and "the social"? In an early essay, written for a Columbia University seminar when he was twenty-five years old, Ambedkar addressed the relationship between social status and sexed scarcity by focusing on the elimination of high-caste "surplus" women through incitement to *sati* (widow immolation). In that essay, *Castes in India*, Ambedkar underscores the artificial nature of scarcity, which is produced by ridding society of widows, surplus women, through resort to ritual injunction for good wifehood. Since the institution of marriage also institutes caste, anxieties about birth, descent, genealogy—sexual intimacy, in brief—will forever stand behind the question, "What is your caste?"

Jaaware does not follow Ambedkar in his desire for origins, even though Ambedkar and his predecessor, the radical *satyashodak* (truth seeker) Jotirao Phule, are recurring presences throughout the book. Institutions require the fiction of origins. So, too, do constitutions. As chairman of the Constitutional Drafting Committee, Ambedkar is modern India's reluctant founding father. In his writings and speeches, Ambedkar draws our attention to the gap between legislated redress and existential suffering, and between the founding document and the equality it imagines. The recurrent staging of these aporias also allows Ambedkar to underscore the impossible yet necessary task of annihilating caste.[2] Instead, Jaaware scales up from the body to literal and metaphorical layers of touch. He calls *Practicing Caste* a tropology and invites us to view caste (as system, structure, grid, or field) as a potent, powerfully precise virtualization of touch. Much rides on the singularity of touch. Touch is both ubiquitous and necessary. Touch, skin, and body function as a threshold between "inside" and "outside," and together they illuminate the generative power of the haptic. Touch has the ability to mediate between empiricism and abstraction, and between the literal, the metaphorical, and the metaphysical.

What I have called the "virtualization of touch" is tracked here with deceptive simplicity. This is philosophical thinking at its best—concrete, everyday, and open to new ways of thinking about old words and worlds. In a powerful discussion of capitalism and religion, Jaaware maps the difference between caste and class onto the distinction between societies of inheritance and acquisition, respectively. In turn this is predicated on how each understands the relationship of intellectual and manual labor, and therefore the business of *abstraction*. However the text's tour de force lies in Jaaware's attention to rhetorical categories (literal, figural) for classifying touch along the axis of good and bad valuations, which is scaled up to consider literary questions of "style," and "genre."

Practicing Caste turns to "Dalit literature" and criticizes efforts to read literature as social document. Elsewhere Jaaware writes, "It seems to me that quite a lot of dalit literature is driven by a well thought out, educated sense of justice and injustice, where equality is repeatedly understood in the capitalistic sense of equal opportunity. . . . Most often, such literature is received as narratives, or lyrics, of suffering."[3] Jaaware is deeply resistant to this politics of identity that has resulted from the institution of Dalits as figures of social suffering in the contemporary political public sphere. Instead, his own effort is to read Dalit literature as *destitute literature*, as writing that stages caste's persistent and brutal dehumanization as occasion for ethical decision. In posing caste as a problem for ethics, Jaaware returns to that fundamental question of what it means to *be-with-others* in a startlingly new manner.

Aniket Jaaware has written a breathtakingly beautiful book whose form and content mirror the provocation to unlearn what we think we know about caste. We would do well to travel a while with a text that has so much to teach us about being together and apart.

Anupama Rao
New York
December 22, 2017

PRACTICING CASTE

INTRODUCTION

Whatever is not truth cannot measure truth precisely. (By comparison, a non-circle [cannot measure] a circle, whose being is something indivisible.) Hence, the intellect, which is not truth, never comprehends truth so precisely that truth cannot be comprehended infinitely more precisely. For the intellect is to truth as [an inscribed] polygon is to [the inscribing] circle. The more angles the inscribed polygon has, the more similar it is to the circle. However, even if the number of its angles is increased ad infinitum, the polygon never becomes equal [to the circle] unless it is resolved into an identity with the circle. Hence, regarding truth, it is evident that we do not know anything other than the following: viz., that we know truth not to be precisely comprehensible as it is. For truth may be likened unto the most absolute necessity (which cannot be either something more or something less than it is), and our intellect may be likened unto possibility. Therefore, the quiddity of things, which is the truth of beings, is unattainable in its purity; though it is sought by all philosophers, it is found by no one as it is. And the more deeply we are instructed in this ignorance, the closer we approach to truth.

—NICOLAUS CUSANUS, *ON LEARNED IGNORANCE*

This book attempts to understand operations of touching and not touching, initially in themselves, but increasingly in their social operations as they relate to caste. The study of caste itself now has a long tradition and is divisible into studies that support caste and its operations, and those that do not. I offer this study of touch and caste in order to understand these operations so that we can find ways to end them.

To arrive at an understanding of caste that is different from those already available in the saturated field of caste studies (and dalit studies), one has to attempt to think *everything* differently; and to do that one has to stop depending on material—texts, ideas, analyses—already available. This is, I think, a weak version of the phenomenologist Husserl's notion of *epoche*, the withholding of belief in the existence or in-existence of the object under study, and the suspension of the natural attitude—a methodological first step in what is often called the first reduction. I explain this in terms of a deliberate forgetting. A significant amount of deliberate and indeliberate forgetting is necessary in this project. However, as many have pointed out, not least Heidegger (and the hermeneutical tradition that begins with his ideas), such a deliberate forgetting is not really possible. Therefore, I will occasionally pretend to forget what cannot be forgotten, keeping in mind that there is an indeliberate forgetting over which we have no control. Allow me to elaborate on this a little.

OUBLIERRING

For some relatively independent thought to occur in one's mind, earlier practices need to be *oublierred*—to coin a portmanteau word from the French *oublier* and the English *err*. This attempt at deliberate forgetting or ignoring is easily misunderstood as *arrogance*,[1] for it sounds, literally, as if the speaker is authorizing himself or herself to say *anything* he or she feels like saying. But this is necessary for some relatively independent thought to occur in one's mind. The thought and its relative independence are more important and urgent than matters of authority, authorization, and agency for that thought, because one can only say *something*—one cannot ever say *anything* (unless we mean the utterance of the word *anything*)—so there is no need to be worried that the author authorizes himself or herself to say *anything*. The *anything* merely gives an abstracted indication of positive possibilities (often taken to be negative) that necessarily concretize or manifest or articulate themselves into finite *somethings*. In short, even if one wishes to say *anything*, the moment it is said, it becomes *something*.

In this world that we inhabit, which for me is an always already changed world, what can we do or think or say that does not remind some reader of something or something that some reader cannot predict or something that simply is foolish?

Such *oublierring* cannot succeed fully: It is an *erring* because both author and readers command, and will remember or recall, earlier instances of related, similar, contrasting, and/or complementary thoughts, some of which, to be sure, the author is not aware.

This book is an attempt to posit an initial condition of possibility, an attempt at *oublierring* whatever I might have known about caste, philosophy, dalit literature, and so on before I started writing this down. I thought it necessary to do so in order to be able to have relatively independent thoughts on caste, society, metaphysics, philosophy, and a few other matters. What one cannot deliberately *oublierr* (but in fact also could be said to be always already *oublierred*) is sometimes called second nature. Because of my basic training in literary studies and the little reading that I have in related areas in the humanities, linguistics, sociology, philosophy, history and so on, such thoughts as occur in the rather airy space of my mind tend to be configured in literary/ rhetorical terms, concretized, in this particular case, in terms of a tropology.

I also argue that given this methodological step, a study of caste can proceed only in a deductive manner, since the operations of touching and not touching are not enumerable, de facto, and, even more strongly, de jure. The first two chapters attempt such an a priori study and understanding of operations of touch.

One effect of *oublierring* gives me another imperative: to think of society and the social operations of touching and not touching, but not in a received "sociological" manner. This imperative leads, after a few chapters, to questioning the categories of "society," "social," and "sociality," that is to say their usefulness to understand operations of caste, as seen in touching and not touching. The third and fourth chapters focus on this aspect of touching and not touching as they relate to caste and society and their history.

As I mentioned earlier, my exploration thus begins with a tropology of touch, expanding into a rhetoric[2] of touch. Later this rhetoric expands into a few other dimensions. The generalizability of some of the observations is tested, and they are found to be generalizable: taking touchability/untouchability *out* of its specific "Indian" context in caste and caste studies.

Given the phenomenological tendency to look for the as-yet-unseen level in the analysis, my attempt is to seek and find at least one level of description *below* the generally available "deep" analysis of "social" and "literary" and other kinds of phenomena, and I believe I have found it once or twice. Having done so, I hope I shall be forgiven for all that is irritating and/or annoying in the text that follows. Be patient dear reader: My goal is, in the language of computers and programming, to understand word-processors, and after passing through a phase of advanced programming languages, to see if I can reach and cognize "assembly" level work, and not lose the ambition to write an altogether different compiler. In other words to look for the lowest level constituents in the constitution of caste.

INVITATIONS

Although it is true that it is just me who is saying it, I should say that the text that follows *invites* readers to think aloud *with* the text, perhaps even think aloud *against* the text. The archaic "we" that I have used throughout the text is (meant to be) an *invitation* to form an "us" (perhaps an "us" that does not need *my* consent, or is formed without my consent, even an "us" without me, because some time or other, I will be dead).

The text also invites you to think in a minimalistic style. It attempts to do this through minimalistic syntax (that does not preclude verbosity or reduplication of diacritical signs and brackets) and what I might call an asceticism of thought and language. This takes away the necessity of an elaborate scholarly apparatus of citation and reference and footnote, de-signed to persuade readers (through a rhetoric of academic presentation of material and argument), that I am or might be right, that such and such texts and the authority of earlier authors back me up.[3]

It has now become clear to me that such *oublierring* is also an intellec-tual risk which precisely lies in the possibility that the style, the thinking and the style of thinking, will be mistaken as precritical and uninformed, and minimalism of citation and mention mistaken as mere name-dropping, that conceptual and stylistic allusions will be missed and read as the author's assertions and so on.

But I realized, as I wrote, that this is what every text risks (knowingly or unknowingly), and on that count there is nothing special about this partic-ular text that follows this introduction, for even with abundance of citation and authorization and even syllogistically presented conclusions one can be mistaken; one can cite and cite, and still be incorrect or mistaken; one can write and write and still not get across; one can speak and speak and still be unheard.

There are other invitations implicit in the presentation of the text, for example the implicit claim in the text (not implicit *here* anymore, but dear reader, you are likely to forget what is explicit *here* by the time you read what is implicit *there*); that more or less traditional, if not orthodox, techniques of literary analysis (like a tropology), might be useful for understanding and describing "social" phenomena (like "caste"). Given the poor condition of the humanities and literary studies across the world almost and given the mediatized success of the misnomered effectivity of social capital and Corporate Social Responsibility, or what goes in the name of "development," this seems an important point. This also implies that the construal of the object of knowledge called "caste" can proceed differently if we do not take available sociological descriptions as always already valid, or as facts, social or otherwise.

MOVEMENT

The text moves through a series of philosophical and methodological flavors, initially a Husserlian phenomenology, later a structuralism, and much later a poststructuralism. On a few occasions I offer a reading of some events, some narrated and some observed, and some literary works. To some extent the text is a response to the already saturated discourse of "caste" and "dalit literature." However, inasmuch as it is also an attempt at thinking anew about these, it does not assume full familiarity with these discourses, though in places where I have failed to see the need to provide glosses, the text might sound obscure. Again, it is necessary to think it anew now, at this point in history, because the problem is now recognized worldwide as a problem of social justice in India, and more and more scholars from outside India are recognizing "dalit" literature, albeit mostly as one more addition to "literatures of protest" and/or "political literature." (I discuss dalit literature in the fifth chapter, which also serves as a kind of transition away from the earlier focus on touching and not touching, toward a more general and conceptual discussion.)

As far as the reception of dalit literature is concerned, the passage from ignorance (of "dalit" literature) to knowledge can happen only through translations into English. Like most other things, this too is both good and bad at the same time: good that this literature is getting into English, and bad that it is getting into English and other dominant languages of the world. Many of these translations will not be patiently done, since publishers must sell this literature while its popularity lasts: The translations will be quickly done and in an English that it is internationally readable (and therefore rather bland, a very small and thin and narrow ribbon in the available bandwidth called the English language). Many of these impatient translations will not have the time to reflect on translation itself; neither will they have the time to pay attention to linguistic detail of either of the two languages. There will be sins of commission and omission.

Inasmuch as "caste" is understood mainly through sociological and anthropological categories, and "dalit" literature is understood as literature of protest against social injustice, consequently receiving more sociological and political than literary attention, any attempt to think anew about "caste" will have to withdraw from most sociological and anthropological understanding.

Therefore the double movement: a simplified phenomenology and a somewhat less simple tropology and rhetoric. The text that follows begins with a description of touch, eventually ending by putting the social (as always already existing) under question, pushing the issue of sociality into a futurity-without-knowledge, a future to which we are necessarily blind, a future in which we are gone and past.

The problem thus runs away from the topic of touchability and untouchability and even "caste." It is my belief that this running away, abandoning, is necessary for the annihilation of caste and the collective reconfiguration of the idea of equality and democracy.

Nevertheless, the observations I make about touchability and untouchability are hopefully interesting, and perhaps some of the analyses useful.

THEMES AND HORIZONS

The major theme is that of "touch," however, "touch" as it constitutes (and is constituted by) touchability/ untouchability (a matter usually understood as part of the more general theme of "caste") in "society." A very large body of work, including that by non-Indian scholars, has been available on the general theme of "caste" in India, mostly within the fields of knowledge called "sociology" and "politics" and "anthropology." In fact, the amount of writing available is large enough for it to prevent me from attempting a "survey of existing works," or even a summary of important arguments as something you should know before you engage with the main text. The horizon is necessarily fuzzy. However, a few remarks on some more recent work might not be inappropriate here.

One text that seemed to depart from existing traditions in the study of caste is *The Cracked Mirror: An Indian Debate on Experience and Theory*, by Gopal Guru and Sundar Sarukkai.[4] Several scholars interested in the issue, or generally in South Asia, are likely to be familiar with the essays in the book. It initiated a very welcome debate on caste and touchability/untouchability as a point of departure for a larger debate on theory and experience and social sciences and social theory. Although it does inform us about various ways in which caste and touch could be thought of, including a phenomenology (in the loosest possible sense in Sarukkai's essay "Phenomenology of Untouchability") and an archaeology (Guru's essay "Archaeology of Untouchability"), the essays in the book seem to take touchability/untouchability as already constituted facts and/or practices. The essays do not seem to explore how these are constituted, in the phenomenological sense.

Yet another very recent text is *Dalit Studies*, edited by Ramnarayan Rawat and K. Satyanarayana, which in a certain sense, announces the final moment of "arrival" within the academy of the eponymous topic.[5] Several essays in this volume seem to desire to break away from earlier ways of studying caste (which is a good sign), while still unable to give up a moral notion of injustice and exploitation (which is not a good sign).

My attempt in the text that follows is almost the opposite of these approaches, although I share the same point of departure more or less. I almost ignore the well-rehearsed arguments about caste as the major signifier of injustice and violence in Indian society, and my attempt to understand ethics as they relate to caste starts from a short story by Baburao Bagul, as seen later in the text (chapter 5), rather than a general discussion of theory and/or ethics. My argument there is simpler: Because storytelling has the ability to terminate the telling at crucial points of ethical or political "destitution," it provides a special opportunity to abandon the narrative at an ethical conundrum or even an aporia, thus allowing us, in fact forcing us, to think about it. In real life, perhaps there is much less time to *think* because it is necessary to *act*, and act *immediately*.

Another somewhat implicit theme is that of "dalit" literature, though it is specifically addressed only much later in the book, and that too only in a few pages. In fact, as I point out at the very end, in the Coda, the exploration, perhaps even an investigation that might be reflected in the main text began while I attempted to understand Marathi "dalit" poetry.[6] This literature is the best-known feature, as of now, of Indian and non-Indian understanding of the relationship between caste and modernity (dalit literature as a fully modern phenomenon). Quite a lot of non-Indian interest comes to "caste" after encountering "dalit" literature. Here again, I cannot offer a survey or a summary. Neither do I engage with texts in much detail, because I think this book is not really the place for offering elaborate readings of some texts.

Yet another consideration that is important for the text that follows is that of "methodology": *How* shall we study caste so as to produce a new knowledge of it? After some thinking it became clear to me that no single and pure and internally consistent "method" was adequate. Therefore, the shift is from initially *a gestural Husserlian phenomenology, later a structuralism, and toward a poststructuralism.* The last I indicate by the term *destitution.* I have elsewhere talked and written about this.[7] I have already indicated that none of these can be found in their purer and stricter forms in the way my argument proceeds.

Another theme that begins in chapter 3 is that of tendencies in societies, and I posit two basic tendencies, that of *inheritance* and that of *acquisition*: In most societies one of these is the prevalent tendency. Things are understood as "inherited" (caste and gender, e.g.) or acquired (knowledge, commodities, wealth, etc., but also gender and caste, and many other things). These are also related to what a society in general sees as the source of value: If value is "inherited," it is not of our making; if it is acquired, it is of our making.

A theme that does not receive great attention initially, but becomes increasingly significant as my argument proceeds, is that of *sociability*, as distinct from *sociality* or, more clearly, from *society*. Perhaps the quickest formulation of what I could conclude is: *There is nothing called society; what perhaps needs to be brought into existence, in our living together, is the possibility of new forms of sociability, even at the risk of the forms becoming transcendental in the Kantian sense.*

Perhaps I should also describe what I have assiduously attempted to avoid, *to oublierr*: (a) the academic automatism that most likely takes academics like me to "classical" texts and sources (the initial argument in chapter 5) as an automatic first step toward understanding caste; (b) what I polemically call a "fake encounter" with injustice that makes us bemoan the injustice of caste and celebrate "narratives of suffering." Such fake moral encounters generate a premature moral satisfaction, happy moral judgment on ourselves, and make us neglect, if not forget, those who were killed in caste violence, those who were raped, or those who died doing caste-related work (the sewage worker who dies of poisonous gases in the sewers of large urban settlements), in short those who could not tell stories of their suffering; and (c) a sociological and/or anthropological reliance on terms and phenomena such as endogamy and exogamy and commensality and connubiality to explain "caste": My argument in the later parts of the text is that these are phenomena at a level much higher than that of touchability/untouchability, these are *in-stitutions*, whereas we need to discuss what happens at the level of what I have termed *de-stitution*. The last two chapters indicate this movement, one section in the very last chapter enacts a kind of linguistic destitution before some workable futures can be discussed. That which began as a weak withholding (from a phenomenological perspective) transforms itself into a destitution.

Less assiduously, I have avoided a detailed discussion of the difference/similarity between "caste" and "class." This has been debated a lot, and again it is not really possible to give an introductory summary or survey. However, I do make some argument *en passant* that inasmuch as societies that are more easily understood as "class-based" also have regimes of touchability and untouchability, there might in fact be, possibly, a different way of understanding the difference/ similarity between caste and class. In brief, caste has been understood in terms of class; perhaps the reverse, considering class in terms of caste, might be useful. Again, this is not explored in any detail, since this debate is not really the central concern.

ENDING REMARKS

I hope to have given some sense of what follows, to have led you into the main text and argument. Ideally, I should have been able to write in a way that requires no prior knowledge of what the book is about. However, it is important to give a brief description of the frame within which the argument operates, which I hope to have given above. Those readers who have some familiarity with caste, or phenomenology, or sociology of caste in India will sometimes find themselves on familiar territory, but sometimes on almost completely unfamiliar ground.

I have already said this earlier: The purpose behind the way the argument is presented—in style and movement—is to invite you to think along with the argument, with as few intellectual accoutrements or paraphernalia as possible. Thinking—adventuresome, and destitute.

1

TOUCH AND ITS ELEMENTS
AND KINDS

PRELIMINARY REMARKS ON TOUCH AND ITS ELEMENTS

We treat here of touch itself, its elements and its kinds.[1] In every inquiry addressed to any of the senses, the specificity of the experience of the senses causes difficulty. Every experience of any of the senses is particular and specific and, in all probability, unrepeatable unless conceptualized and classified. It is possible to enter into a traditional metaphysical discussion regarding this particularity by stating that every and any experience becomes an experience through the synthesis of sense-data and mental categories: These categories exist before experience, since human beings come preloaded with them, though the categories come into operation perhaps only at the moment of synthesis.[2] It is also possible to understand these mental categories differently, as precomprehensions that might subsequently get modified into the question of being.[3] These problems can be discussed in a variety of ways. However, we will not undertake a discussion of these problems, for the following reasons.

Our intention here is twofold. It is first to use touch as a double-edged blade that will cut through, on the one hand, philosophical-metaphysical difficulties and, on the other hand, sociality itself, which I understand as the social issue of caste. In short, we intend here to anatomize philosophical-metaphysical issues by an analysis of the sociality of touch, and in turn, anatomize the sociology and anthropology of caste by a philosophical discussion of touch. Second, the intention is to understand society and culture—the world as such—through the analytical prism of touch and see the spectra and colors and radiations of which these are composed and their diffractions and dispersal. But our intentions never have sufficient force to avoid issues of synthesis, or of precomprehension, or epistemology or philosophy. There is a stronger reason as well, which is that touch is a *material* phenomenon, neither easily susceptible to an idealization that is essential to most philosophical discussions nor easily intelligible without such an idealization. A discussion of the materiality of touch generates its own risks and its own contradictions. We use the term *materiality of touch* in its most physical and intimate sense of touching mother, father, lover, and friend, touching one's own body, touching things, and so on, touching as such. We suggest that one's body is more intimate to one than one's mind. In suggesting that the body is more intimate than the mind, we do not merely have the dubious profit of inverting a dominant tradition but also the benefit of having to think about the body itself as a new problem (assuming that intimacy is not the same as knowledge). What is new about the problem is that, unlike in the life sciences and the physical sciences, we will not be able to ground knowledge of the body in a purely static and mechanical conception of materiality because we are to think of the materiality of the body in social terms and not merely as a unit of condensed matter: plasma, blood, bones, cells, and the mitochondria in them, and so on.[4] We use the word *body* in its simple sense. Later on we discuss the genesis of the body itself.

Also what is new in this problem is that we are approaching the body from the point of view of touch (and not any other sense), and this is appropriate because touch is something that the complete and extravagant surface of the body is capable of sensing (other senses have specific locations in the body: the eyes, the ears, the nostrils, the tongue—all located very close to each other in the head). It becomes clear from this primary observation that it would be misleading to understand the body on the model of any one or all of the other senses, since they are located in specific parts of the body and therefore can give only a partial model of the body. Another reason is that even if we accept the position that mental categories work on (from this point of view always already hypothetical) "raw" sense-data, the position does not subtract anything from the particularity and materiality of the "experience" itself. The notion of materiality has that added advantage of precluding the dichotomy between "real" and "ideal" objects. These reasons we take to be temporarily sufficient for suspending, if not foreclosing, the metaphysical/philosophical problems of synthesis and precomprehension, or their various equivalents.

Although it might seem wiser, as well as methodologically more sound, to locate touch within the general ensemble of senses, We intend to discuss touch itself, partly as respectful homage to phenomenology (Husserlian phenomenological protocols, as well as Merleau-Ponty's *Phenomenology of Perception*).[5] There is yet another reason as well, which is that, quantitatively, touch is the largest sense of all, and thus it is more diffused, dispersed, and unlocalized than any of the other senses. Touch has no specific locality or home in the body. Within the body, touch is homeless: It traverses the body; it is available on all the planes of the body. It needs to be pointed out that those locations where the surface of the body twists, folds inward, or is invaginated become primary locations for our sense of an "inside": the mouth (the same continuous plane, actually, as that of the forehead, the nose, the arm), the "inside" of nostrils (again the same continuous plane), vaginas, anuses, ears, in short the "inside" of the body is actually the same continuous surface as the "outside." It is also to be noted that organs in which the other senses are located are capable of touch as well. Parts of them are extremely sensitive, as we discover when something gets into our eye. This does not only mean that these organs are more versatile but also that touch is the primary sense, since it is available on the whole plane of the body. Touch is present wherever there are nerves, and nerves are everywhere.

In order to be able to discuss touch at all, we need to isolate the sense and experience of touch and discuss its particular characteristics, distinguishing it from other senses. These distinctions among the senses are extremely important, for these will later allow us to assemble something that we call the body. (The body-as-a-whole is structured by the various permutations and combinations of these distinctions.) We have partly indicated a few of these characteristics; however, a more detailed description is necessary. We identify the following as the elements of touch.

Physical Elements

As we have already stated, a fundamental characteristic of the sense of touch is its materiality. Apart from whatever advantages there might be in the metaphorical meaning of materiality (especially from a Marxist point of view), we are equally interested in the more physical—and therefore the perhaps literal—meaning of materiality. From this point of view, it becomes essential to define clearly the physical elements of touch. We must make sure that we do not quickly understand the particularity of touch as stemming from the individuality of a self which is thought to possess a body, and in that possession, is transcendent to the body (or even transcendental). Although we speak of places on the body and so on, we do not talk of the individual or the "embodied" self. The locations of the various senses, we suggest, are subindividual phenomena: The individual embodied self can be understood to be a macrolevel collection, probabilistically held together. The four major physical elements of the sense of touch are as follows.

1. *Inertia.* The sense of touch depends on contact between skin and something else (inanimate object or texture or animate skin). The object touched must resist the movement of skin against it for there to be contact between skin and object. This resistance need not be a willful resistance—the physical-material properties of all objects, including animate skin, ensure that there will be resistance, which is traditionally explained by the concept of inertia.

2. *Density (and extension).* Similar to the thresholds or limits operative in other senses (twenty-four frames per second for eyesight; 20 to 20,000 Hz for hearing, etc.), there is a threshold of density to the sense of touch, though there are no commonly available measurements. If objects have a density less than this threshold limit, one cannot touch them (in the case of wind, the threshold of density still operates: The rate at which particles strike the skin must be proportionate to the minimum sensible density of objects for us to be able to feel the wind on the skin).[6] It is also important to remember that inasmuch as every sense is physical, in all the senses of the word, all the senses are dependent on materiality. The sense of hearing, which is perhaps the most spiritual and hence most susceptible to idealization, also is material: There are material sound waves striking the eardrum. The difference, however, is in the density: Sound waves are less dense than "solid" objects. The case of eyesight is similar. We consider density thresholds to be clinching factors in determining the relative susceptibility to idealization of the material senses.

Similarly, we cannot touch a thing that does not have any extension whatsoever. One cannot touch points or lines when they are conceived geometrically. One can, in reality, touch only surfaces, which by definition are extended bodies. Further discussion of extension need not be undertaken, for all extended bodies are dense (i.e., have a certain density), and we have already discussed density as it becomes relevant to touch.

3. *Reality.* There are no fictive touches, although the sense of touch might be active in dreams. It would be methodologically unsound to conflate dreams and fiction, and therefore it needs to be stated clearly that touch possesses a reality that is undeniable and not open to fictionalization. The sense of touch is the most difficult to deceive. It is therefore not surprising that there are no art forms that utilize the sense of touch fundamentally. It is tempting to think of sculpture, with its vocabulary of textures and planes, as a touch-dependent art form; however, one is rarely supposed to, or allowed actually to, touch pieces of sculpture.[7] The only human activity in which the sense of touch is given full expression is that of making love, and only that could be seen as an art form dependent on touch. However, it would be entirely misleading to think of it as an art form. (It should be noted though, that there are no art forms that fundamentally utilize the sense of smell or the sense of taste either: The skin, the nose, and the tongue are the least fictive and artistic organs, rooted in reality as they are.) One might object, pointing out that there is touch in cinema, and even more tangibly in drama. This is misleading again. Touch in drama becomes relevant within the general interaction between actors, since it is relevant and available to the audience only as a visual experience of touch. Touch in drama is not really a matter of touch; it is a matter of seeing some people touch each other. This is not to deny that actors or actresses kissing one another in a play might actually experience touch and possibly sexual arousal. This would be of no concern to the audience except as curiosity or voyeurism or vicarious pleasure. In any case, the experience must be rare. From this point of view, there is no difference between drama and film; both employ visual images as their main mode of expression.

It could be said that the word *reality* has philosophical implications that we need to consider before we can talk of the reality of touch. Our very simple argument on this is that there are very few words that do not have philosophical implications, and we use words as we can and think fit. It seems to us, moreover, that our sense of materiality too would be compromised if we were to use some other word. We are looking for the solidity, stability, and reality, the fleshy thickness, that ground the sense of touch.

Another preliminary observation on the differences and similarities between touch and other senses will help clarify this. Eyesight can be stopped, "shut off" by closing eyes (one's own, as much as others'). One can stop one's ears or one's nose in some circumstances of cacophony or its olfactory equivalent. One can refuse to open the mouth in order to refuse to taste (though this is different from closing eyes or shutting off ears). But a way of shutting off the sense of touch is not available.[8] As long as one is awake, one possesses the sense of touch. One cannot control or reject the inputs of the sense of touch (the more physical forms of torture and eroticism utilize this inability). If someone cannot be touched awake, shaken awake, then he or she is likely to be unconscious.

4. *Contact.* This is the most obvious physical element; in fact, it is the constitutive element of touch. Here we are thinking of contact in *quantitative* terms. If one is touched simultaneously on the various planes of the body, there is a sense of envelopment or cocooning, which is accompanied by the most intense emotions, whereas minimal touch can be made into a serious and aggressive form of insult, especially when more touch is expected while touching human beings. One finds parallels to this in touching inanimate objects as well.

In short, the philosophical argument in the above enumeration of the physical elements of touch is that the metaphysical is an allegory of the physical, and not vice versa, which is how tradition, especially religious tradition has it.[9]

Nonphysical Elements

Materialism, even when emphasizing the physical aspects, need not fall into an insistent mechanical empiricism of the body (ultimately resulting in naïve behaviorism). Moreover, as already stated, my interest is, finally, in the social implications of touch, and although sociality can be understood materially (there would, in fact, be no other way to study society without falling into a naïve behaviorism or into some form of idealism), it cannot be reduced to questions of empirical individuals who could be said to be capable of stimulus-response interactions.

1. *Repetition and attention.* The number of times touch is sensed is an important element. One's sense of touch on the buttocks when one sits or lies down is something to which the body does not pay too much attention. This is, in all probability, a result of repetition. However, it must be noted that not all repetitions are relegated to the background: The sense of touch

in fingertips—one of the important places of touch—is paid more attention, and the same is true of lips, and, obviously, the various "insides" of the body. Thus, one could say that although the whole body is the organ of touch, there are some locations on the body to which more attention is paid. Attention is also a matter of activation: One could equally say that when one wants to touch something, nerves in some places of the body are activated first—notably fingers and the palm in general. We have hands to grip, but before gripping, to sense by touch.

2. *Emotion.* There is no touch that is not accompanied by some emotional charge. There is a kind of touch we could call the "dead" touch, but what is notable is that the "deadness" of the touch is itself an emotion. Other instances of emotional touch are obvious and need not be enumerated; however, we might point out that primary forms of pain and pleasure are entirely dependent on touch. The secondary, more advanced, forms utilize other senses. The special relationship between touch and pain needs a thorough investigation, which we are not able to undertake here.

3. *Sociality.* Sociality is an important element for our purposes. It is through the prism of the notion that touch is a social phenomenon that we are to look at society and the hierarchies within it. Thus we begin to think of the sociality of touch by stating that there is no touch that is not social. Prima facie, it looks as if touching oneself—especially one's "intimate/private" body parts, is a nonsocial, purely personal act. A close look shows that this is not the case: Touching oneself is as social as touching others. There are no degrees to sociality: Touching oneself is as social as touching members of the family—mother, father, brother, sister, child—or touching people relatively less than "family," friends, or strangers.

4. *Intimacy/proximity.* There are degrees, however, to the "intimacy" associated with touch, and there are degrees to the intensity of emotions attached to touch. Later on, we make a distinction between touching oneself and touching others, but it should be stated clearly that this distinction is based on the degrees of intimacy and emotion. Each touch is accompanied by a sense of intimacy, minimal or maximal. The notion of intimacy is deeply linked to the notion or sense of proximity. What we can touch easily, with minimal bodily expenditure of energy, is the most proximate, and that which is most proximate is likely to be understood as the most intimate. It is this proximity that generates a sense of territorial possession. Our understanding of an alleged near-absolute possession of our body is premised on

the fact that we can touch ourselves without much expenditure of energy and with few social restrictions. It is clear that the varying degrees of the senses of closeness and distance in social relations are dependent on whom we can touch without much expenditure of physical or mental energy or, for that matter, social energy.

Having enumerated the elements of touch, we continue with the description of the general features of the sense of touch, based on preliminary observations. These observations run the risk of being grounded in a conception of materiality that we have already castigated: the scientific conception of materiality. The discussion of the physical elements of touch too ran the risk. Before we begin a discussion of touch based on some more of preliminary observations, we need to make explicit the difference between a scientistic and naïvely empiricist notion of materiality and observations that are reminiscent of, or residually are, phenomenological. We say "reminiscent" and "residual," for these observations are not phenomenological, strictly speaking. We have not performed even the first phenomenological reduction, although we have isolated touch and chosen it to be the prism through which to look at the world. Neither have we performed the second reduction, the transcendental-eidetic one, though we speak of the elements of touch, which are, a priori, and thus necessarily, *essentially*, to be found wherever touch is to be found. We will first clarify the difference between the phenomenologically sound procedure and the procedure that we have adopted. The difference lies in the fact that we have placed some value on the materiality and *reality* of touch: This would be impossible in strictly phenomenological protocols, which would need that the belief in the reality and/or unreality of the observed phenomenon be suspended in the first reduction itself.

Our procedure is different, since, as already stated, our interest is in the sociality of touch. This also clarifies the difference between the scientific notion of materiality and our own conception. In the scientific conception, materiality is reduced to matter (thus causing great difficulty in defining the difference between the animate and the inanimate, between human and artificial intelligence, resulting in the great gulf between science and ethics). Thus the neurosciences might give us more or less exact descriptions of neurons moving across synapses, and firing or not firing, but still fail to predict, or even explain fully, the phenomenon of emotion—especially its social nature. This does not mean, naturally, that the kind of knowledge of touch that might be produced on the basis of such a conception of materiality is, or would be, invalid or useless. It might always be possible to alter mental states by psychotropic chemicals, for example, to the point of inducing or suppressing paranoia, but that is not, clearly, an area within which social ethics can be identified or practiced or, for that matter, evaluated. Science can progress scientifically only by pretending that matter is a *natural* phenomenon. In this sense, science is the most Romantic of Western developments, one in which everything is given or is believed to have, or be capable of, a natural explanation. Quite a lot of the characteristics of Romantic lyric poetry, in fact, can be found in the scientific conception of materiality, in which material experience is an isolated phenomenon, usually of an interaction between a fully formed subject and a strikingly "mysterious" object, the concentrated and emotionally charged "gazing on the object" (usually in solitude), and perhaps, above all, the conception of the notion of the "experiencing" (experimenting) self.

Our concern, however, is with the sociality of touch. It is important to clarify that it is the sociality of touch and not "the social construction of touch," the latter being a Romantic, "constructivist" interpretation of things. There is, substantially very little difference between saying that a subject constructs phenomena and that phenomena are socially constructed. For sociality to be fully appreciated, more than lip service will have to be paid to those who are other than ourselves, including those times and places when we are not ourselves. Moreover, quite a few of the conceptions of society seem to be conceptions of a self/subject that is merely multiplied several times over, a kind of demographically gigantic subject, just as quite often the conception of others in the self/other pair (it would be entirely wrong to put the latter in a capitalized or singular form) is again a conception of the self superimposed or multiplied (with positive or negative signs) onto a rather vague and somewhat poetic notion of alterity. My interest, however, is in the social prose of the world, especially how it gets written, by touching oneself, or touching others.

(Let us offer an explanation, or justification, of my use of "scientific" sounding words: *density, inertia, surfaces, planes,* and so on. It should be noted that at no point have we mentioned any form of countability or accountability or calculability while discussing these. While talking of feeling the wind on the skin, we talked of the rate at which particles strike the skin being proportionate to the least density that is touchable. Though this actually might allow one to physically calculate the thresholds for the sense of touch, we speak of the rate only to demonstrate that what looks like "fictive" or "false" touch is only a result of the proportion between spatial density and temporal frequency at which particles of air touch the skin. At the same time, we must acknowledge that historically, at the end of a century or two of scientific advance, materiality can be expressed and understood only in a vocabulary mostly scientific or apposite to it. There seem to be available only the scientific words to speak of bodily experience or of the five senses.)

Another preliminary observation is that, topologically, one can touch only surfaces or planes. Touch does not have a sense of inside. One cannot touch the inner surfaces of a closed hollow sphere without cutting it open. (Though cutting is, without doubt, dependent on the sense of touch, cutting as such is an action different from touching; it is a different discourse; the point is, one cannot touch insides without leaving touch itself, even if for a moment.) The cut destroys the sense of the mysterious inside of the hollow sphere because it renders the inner planes more or less continuous with the outside plane. Naturally, this depends on the thickness of the material of which the sphere is made. If it is made of thick material, then one may discern another plane, marked as a separate surface by two "edges" between the outside and the inside. It follows, if we accept the assumption that touch is a primary sense, that the body is capable of touching only outsides of things. The body, of necessity, has this limitation that it can sense by touch only the outside of things. This is proved by the supplementary observation that if we can touch something that is an inside, it can only have been an inside sometime in the past. The inside is, at the time of touching, only a memory. We could say, thus, that all talk of "the inside," and all insides themselves and in-themselves, are based on memory: a mental thing. That which is inside cannot be touched.

This allows us to distinguish touch from the other senses, since though we cannot see, smell, touch, or taste anything that is inside, we can hear something that is inside. We could go so far as to say that only the sense of hearing is capable of postulating an inside. Insides can only be heard, like the speech or silence of someone else's love for you, or Soul, or God, or thought itself. All we have to do is to shake the matchbox to hear if there are matchsticks inside—provided they produce sound within the 20 to 20,000 Hz thresholds.

PRELIMINARY REMARKS ON TOUCH AND ITS KINDS

Inasmuch as the above are *elements* of touch, they are common to all touch and do not help us posit *kinds* of touch. The elements are constituents of the materiality of touch, its material content. Touch, however, has only one *form*, which is that of contact. It can be seen, we believe, that contact itself is of a twofold nature: It is the *form* of touch and, at the same time, the *content*. However, unlike other phenomena that are differentiated on the basis of form, here the form does not allow us to posit kinds of touch on the basis of the contact-form. A formal classification of touch seems impossible a priori. We are therefore forced to look for other principles if we wish to posit—which is to say, differentiate among—kinds of touch. It seems that semantics, especially a social semantics of touch, provides a useful principle. We can distinguish between this and that kind of touch on the basis of the meanings of this or that kind of touch. The description of kinds of touch that follows is, in short, an elementary semantics of touch, the nonphysical material content of touch and contact. A closer look at the semantics of touch quickly reveals that there is a literal meaning of touch and there is a figural meaning of touch.

The literal meanings of touch are clearly discernible when we touch something in order to touch that thing. Of the figural meanings, three are clearly the most frequent: metonymy, allegory, and metaphor. It should be borne in mind that we are building up analogies between the realms of touch and the realm of language and rhetoric (not semiotics, though that too should be possible), and it seems inevitable that the boundaries between figures would get a little fuzzy. It is possible to confuse figures in the realm of touch. The three kinds of figural touch that we are attempting to establish are not based on forms of touch but on contents of touch. All three have the same form, that is, contact. What differs from figure to figure is the meaning of touch, and through that, possibly, the object that is touched. This is most clear in allegorical touch. To touch the feet of elders, or savants, or an idol of god is to touch the realm of the sacred; this is a clear case of allegory. Metonymic and metaphoric touch are most clearly seen in fetishism of a variety of kinds, from leather to foot to hair to lingerie, and other more harmful kinds of fetishism. Condensation and displacement are most intensely felt in the realms of sexual touch.

Along with figurality, there are also the meanings of good and bad touch, and these distinctions exist only in combination with the earlier distinction between literal and figural touch. Thus one can have a good literal touch, as in a mother caressing a baby, lovers caressing each other. That this touch is literal is clearly the fact because touching itself is the meaning of this kind of touch. That this is a good touch is clear enough.

The bad literal touch is more often experienced by women than men. An example is a man touching a woman in a crowded bus. Rape and other forms of physical violations are the most extreme form of the bad literal touch.

One can have good figural touch and bad figural touch. The good figural touch tends to be metaphorical or allegorical, as in touching the feet of the elders (which is metaphorically indicative of "respect," although the objects touched—feet—are synecdochic), or touching the feet of an idol or god, or the threshold of the temple, or the stage before going up on it as a performer (allegorically touching the realm of the "sacred"). The existence of metaphorical and allegorical touch needs some thought, since in the realm of touch, which in its materiality is a realm of extension, one would have thought only metonymies could exist.

The bad figural touch is the basis of caste distinctions and, to a certain extent, class distinctions as well. In the case of caste, it is the shadow of the dalit, or various other metonymies and synecdoches. One might think that untouchability is a practice within Hindu society in general, but there is ample evidence that caste distinctions are maintained even within groups of people who converted to Islam or Christianity or Buddhism. In the case of class, the distinctions are less apparent, although it can be clearly observed that members of the upper classes rarely touch—in fact go to great lengths to avoid the touch of—members of the lower classes. This would allow us to state that contrary to impressions, beliefs, or opinions, caste distinctions are operative in societies other than those that are overtly marked by castes and their hierarchies, including advanced capitalist societies.

Another pair of notions is generated by touch, that of touching oneself and touching others. We have already indicated that touching oneself is as social a phenomenon as touching others, the only difference being the degrees of intimacy and emotions attached to or accompanied by these kinds of touch.

Touching oneself too has its good and bad meanings, its literal and figural meanings. Touching certain parts of the body is thought to be good or bad depending on the meanings of the touch—one touches most of one's bodily surfaces while bathing, but that is a good, hygienic literal touch. One touches one's crotch sometimes in public, and that is "bad" literal touch. There are worse forms of the bad literal touch, naturally. Touching oneself figuratively is touching one's "insides," as when assuring someone of the truth of one's speech, one might touch that surface of the body under which the heart is located. This is presumably a metonymic touch. Masturbation provides an example of metaphorical/ allegorical touch. Generally, there are social regulations on what parts of one's own body one can touch (even when one is alone).

Touching others (and at this point we are using the word *others* with its most naïve content possible, since the question of others and otherness-as-such is to be generated on the basis of the sense of touch): Socially, this is the most significant kind of touch, inasmuch as caste manifests itself as regulations on touching others, literally and figurally, in conjunction with the notions of good and bad. These notions, as is amply clear in the social practice of these regulations on touch, primarily constitute caste. The dalit's touch is by definition bad—if you are yourself an upper-caste.

Lest some people delude themselves on this issue, with the help of the concept of class, as opposed to that of caste, we would remind them that one does not touch the waiter who serves food in the restaurant or the doorkeeper who holds the door open or the cab driver, whose head might come quite close to ours as he turns the meter, or the garbage collector, who collects those black trash bags full of dirty things, or, for that matter, the TV repairer, who, after all, charges "so much" for doing his or her job, or the beggar on the street (even if he or she is a music student playing the violin to earn his or her education). One could make a list here that is several pages long. Certainly one does not touch the man lying on the footpath if he looks poor and drunk. But if he looks well off, perhaps we just may consider checking his wallet for his address, still not touching him as far as possible, and put him in a cab at the slightest hint of consciousness? (Whereas, if he looks wealthy, a member of the lumpenproletariat just might check his wallet for the money?) No doubt the rhetoric of clothes is determinative of our touching him or not, as indeed, gender would be too, but then the class-dependent regulations on touching are amply clear in this example.

Following Maharashtrian anti-caste struggles since the nineteenth century, we take regulations on touch as the mark of caste, and the struggle against "untouchability" (in our description no longer an "Indian" folly or mark of "backwardness") as the prime struggle in various caste reforms.[10] We are of the considered opinion that similar struggles against untouchability, and by extension caste, will give a new direction to leftist struggles as well.

Let us dwell a little more on this distinction between caste and class but continue also to look at it from the point of view of touch. In the context of the class-based hierarchical division of society, what would be the equivalent of the struggle against untouchability? It would mean, for example, that those places to which certain social groups are forbidden access (as Hindu temples were, and in some cases are, for the dalits) would have to open up and admit the lower classes. It is here that the difference between caste and class begins to be clear. In a class-based interpretation of regulations of access and touchability, one would be forced to argue that if the lower classes are forbidden to enter certain restaurants, or shops, and so on, then that is done purely on an economic basis, anybody-with-enough-money has access to these places. The point of course is that even if the regulation were "purely" economic (a purity rarely pure enough), this anybody-with-enough-money is a purely hypothetical category, with the power and ability to actualize it resting in the hands of people of a certain class.

From this point of view one might see the temple-like qualities of expensive restaurants or other places of "conspicuous consumption" (to use Thorstein Veblen's phrase)[11] and wonder that the display of the menu beside the door, so aesthetically printed, does not actually clearly say "make sure that you can occupy, at this moment, the blank economic slot of anybody-with-enough-money." In the case of shops and restaurants the regulations tend to be economic; in the case of Hindu temples (and early Christian churches in India), they tend to be symbolic (to use Pierre Bourdieu's word);[12] but the phenomenon is that of caste. It would be difficult indeed to convince people that a person without money should be given admittance to restaurants and shops as a legitimate consumer/customer. Nobody thinks such inadmissibility wrong, and it is here that the "symbolic" values of money begin to reveal themselves. It is to be noted that for the high castes, the dalit entering a temple, or drawing water from the same well, is as outrageous as a person without money to pay for food insisting on being served in a restaurant as his or her constitutional right. Because the "purely" economic nature of this social organization is constantly punctured in many places by "symbolic" values, it becomes possible to give some interpretation to Walter Benjamin's notion, fleetingly written down in a small fragment, that capitalism is a religion without scripture.[13] Since most of us have more or less given up the economic determinism of orthodox Marxism, it should be possible to concentrate on the manifestations of regulations on touch.

Some of the preliminary kinds of touch that can be posited on the basis of the meanings of touch have been discussed—it is precisely these meanings that regulate (in the good as also the bad sense of the word) our relations with others and our relations with ourselves. That these regulations exist needs further elaboration. We could say that what is being regulated is the material bodily behavior of people. Inasmuch as caste distinctions express themselves through regulations of touch, caste represents the material regulation of the bodies of people. We have already attempted to show that this regulation is operative in all societies, though the supporting rationalization, or ideology, or inner conceptual and emotional articulation might be in one case "purely economic" and in another "purely symbolic" and in yet another a mixture of these two or some other factors. In many societies, except perhaps one in which these regulations are thought to be based on "purely economic" notions, these are firmly attached to notions of the "sacred" and the "profane" or "purity" and "contamination," and "hierarchy." The sociologist Louis Dumont's attempt to understand caste in Hindu society with the help of these concepts is well known,[14] as are the anthropologist Mary Douglas's writings, most famously, *Purity and Danger*,[15] although this book is not about caste.

We differ from the above in our consideration of the kinds of touch, which allows us to suggest that unlike what these authors seem to think, the realms of "sacred and profane" (which in most writings are taken as, and as if, given), are socially and historically generated by regulations on the bodily behavior of people. These two realms and their various mutations, permutations, combinations, and mixtures need not be thought of as a priori categories that exist before society. Regulations on bodily behavior, especially touch, generate these realms in material social practice.

Regarding untouchability as a mark of caste, another observation is important. Untouchability itself is of two kinds. One is untouchable either because (1) one is inhabiting the realm of the "pure" or the "sacred" or because (2) one is an agent of the "bad" touch, the contaminating touch. An example of the first is to be found when a brahman priest is performing a ritual; an example of the other is the dalit's touch. There is considerable social variation in what is regarded as the pure or sacred realm, as there is in who is thought of as the contaminating agent, but these two categories can be found wherever there are caste distinctions based on a division of the social world into pure/purifying and contaminated/contaminating. Notions of activity and passivity interact in this distinction. If one is untouchable because one is of the first type of untouchable, then one can touch others to bless them or purify them. If someone else touches you while you are in that space of "purity/untouchability," then that touch is contaminating (even if it is another brahman who touched you). It would seem then, that if the agency of the touch is with someone else, then that touch is, or could be, contaminating. This would go some way to understand, if not to explain, the general upper-caste sentiment that the dalits are aggressors. The issue, we believe, is of retaining the agency of social touch with oneself, and by extension, in one's own caste. If one is passive when one experiences touch, then it is likely that the touch would be found contaminating. These associations of touch with activity and passivity have several repercussions, several of them indicative of the metaphysical grounding for social practices. We shall discuss these when they become relevant to my discussion of touch. After these preliminary remarks, touch and its sociality can be examined in some detail.

THE GENESIS OF THE BODY

The tiny bundle of emotion and bodily senses that is born does not have a body. To be a possessor of the body is also to be the possessor of the sense of possessing a body; it is as much a matter of consciousness as of the body. In what follows, we describe how the consciousness of possessing a body comes about in the long and arduous journey from neonate to child. This is, so to say, a bildungsroman of the body. We neglect here the parallel, allegorical bildungsroman of the development of consciousness, in keeping with what has been discussed earlier.

To begin, does the neonate possess anything at all? Inasmuch as the traditional description of possession as always already premised on a consciousness of possession has some truth in it, it would be difficult to state that the neonate possesses anything at all. There is, no doubt, considerable difficulty in describing what the neonate is, or does, or says (we are using the ancient Aristotelian distinctions among being, doing, and speaking, indicated, inter alia, in *The Poetics*). The inability to describe the neonate forces us to superimpose on it some pure naturalness of existence, making us think of it either as a primitive organism or as a sadly lost plenitude of being or as a pure automaton (the last in the manner of Renaissance descriptions of animals). One might say, like Heidegger on animals, that the neonate is sheltered in being.[16]

Even within the boundaries of this bildungsroman, concentrating on touch helps us understand the neonate differently. The neonate does not have a body; neither does it have body parts. The neonate's earliest experience of touch is through lips, as they firmly enclose the nipple, and its palms and fingers as they grip the mother's breast. It also experiences the mother's moist lips as she kisses it on the cheeks, or on the forehead, and indeed, on other areas of the surface of its body, for example, its fingers, its belly, and so on. It also experiences what we have called envelopment, when mother and father envelop the neonate in their arms and torso. There is also a potentially unnerving experience associated with this envelopment, that of being lifted up and carried, an experience of weightlessness, of zero gravity, as it were. This is an important experience, since very soon, lifting the baby up is the best, the most effective, and the cruelest way of controlling its voluntary movements. The regulation of bodily movement begins in this stage, and being picked up is one of the first disciplinary actions which one undergoes. A little later, mother and father teach it touch proper. They teach it rather like Saint Augustine was taught language, by taking its fingers and making it touch certain surfaces of its body, as well as surfaces of their own bodies. This is a language of touch made up entirely of nouns, reminiscent of Wittgenstein's account of this type of language. "Nose!" "Eyes!" "Mouth!" It is to be observed, we believe, that the neonate never "touches itself." Mother and father and brother and sister—these are the people who teach the neonate to touch itself.[17] They also teach it, in the act of touching body surfaces, that it is a physical body with extension, with places that it can touch and know as distinct places. Initially, then, the neonate has only these places, not quite a body yet. It is also made to understand that these places are different from the place where it gets its nourishment and the body-enveloping place from which mother and father give it love. In this experience, the neonate also learns, performatively, that the fingers have a pointing function. Deixis of all kinds is planted now onto the fingers, and it will prove to be powerful enough to make us think of words as pointers, as some visual data as pointers, and so on. A lot of our understanding of language, and by extension, semiotics, is based on this pointing function. It is to be noted that in Trần Đức Thảo description of the genesis of consciousness, the pointing function (indicative gesture) has a fundamental role to play, with this difference, however, that it is, for the most part of prehominid and hominid phases of evolution, an inclusive function, rather than an isolating one.[18] It seems to us that in the context of the bildungsroman that we are writing, that is, in the context of the neonate's development of the body and its sense, the pointing function isolates certain places as locations of objects whose names are taught to it by mother and father and brother and sister and even a visiting stranger: "Nose!" "Eyes!" "Mouth!"

These places become a part of the neonate's consciousness even before it develops a sense of the body or a sense of the self. This period is, we should note, the period of the "mirror stage"[19] and the ensuing isolation of a rudimentary self and its sense. We would argue that it is not so much the sense of being isolated from the mother or the ability to identify its own bodily image that gives the neonate its rudimentary sense of self, as the ability to slowly orchestrate these places into a sense of togetherness. Consciousness of these places, their relative autonomy, and their independence becomes more and more refined as the neonate's motor control increases. The increased control also enables it to control the pointing function more precisely, and this in turn gives positive feedback to consciousness, in terms of a more refined understanding and consciousness of these places, like eyes, nose, tummy. Slowly these places are uniquely fixed onto the kinds of data received, thus constituting "organs." There's always mother and father to help in this constitution, with a constant demand that this constitution be undertaken forthwith: "What is this?" "Where is mamma?" "Where's dada?" and answers to these repetitious, inane, but fundamental questions. We believe the visual sense gets "organized" first. Before pointing, the baby looks in the general direction. This "organization" is an extremely important event in the child's bildungsroman. In almost no time indeed, the child will begin to possess organs of sense and learn to make use of them. This constitutes the transition from the neonate stage to the baby stage and subsequent stages.

The ability to possess and guide and manipulate in general these organs of sense generates a reflective consciousness, which is the location of the sense of selfhood. At the early stage, the sense of self consists mainly in being able to manipulate the organs of sense and maintain them in togetherness and relative autonomy. Lacking reflective consciousness, it would be impossible to know if the face that one is gazing upon is the same face, object, thing, or person, whose moist parts "kiss" us (i.e., touch us) all the time.

Along with the reflective consciousness, which has an invaginated structure, to use a medical term, there also develops the sense of location and spatiality; the child no longer attempts to grip something far away but large in size, thinking it is close by. This understanding of spatiality, in terms of what one can and cannot touch by extending one's arms helps the child understand the exterior limits of its organs and isolate and define the space in which these organs are housed. Once these limits are fully established and internalized, now the places that the neonate knew only as places with names, only the nominal stems of touch, and the organs located at these places, come to be enveloped within a body, whose surfaces can now be gazed on directly or indirectly in the mirror. The system is more or less complete at this stage. We attempted to map notions of figures onto the realm of touch earlier, and it seems that the dominant figure in this stage is that of synaesthesia, and the name for this stage of the bildungsroman that is already coming to an end is "Loss of Synaesthesia."

Considerable mental energy is spent in learning to coordinate, organize, and manipulate the senses. The human organism learns these quickly enough. It could. be suggested that the reflective consciousness that coordinates, organizes, and manipulates the senses is the location of the sense of self. One becomes oneself by learning that one has a body-with-parts, as much as by being traumatically weaned away from the mother's breast, or by learning that one is a fragment of the essence of humanity, or by giving the same responses to the same stimuli, or by seeing one's image in the mirror to realize one's isolated unitariness in a burst of joy. No doubt these have their own roles to play, but all these processes of learning and being have to be housed somewhere. It seems to us that we do not really have to look for fictional, imaginary, mysterious, or transcendental agencies because something, that is, the body, is close by.[20]

Once the neonate transforms itself into a unitary body-with-parts, into a child, a number of social regulations begin to be applied to its body. The child is clearly instructed on which surfaces of the body it can touch and which it cannot. The bodily structure of bones and muscles too exercises certain guidelines (by this time the organism has lost the flexibility of joints and muscles that used to allow it to suck its own toes). Arms and legs can be extended or retracted, legs can be used for walking. The palm and fingers at the end of arms become the prime locations of the sense of touch, and the child slowly begins to ignore touch on other surfaces of the body, most notably, the skin over the bottom. The child now cannot touch its genitals in public; the child has also learned that these body parts have to be covered up in cloth. A certain code of touchable and untouchable surfaces of its own body is brought into existence: the genitals, the insides of nostrils, ears, mouth, armpits, and so on. This is accompanied by a general training in deportment, which too is a matter of regulations on bodily behavior. It is from here that the child will begin to train its body in a certain way, eventually through physical training or drill in school; the child's body will become a disciplined topos on which several regulations are always effectively found. The physical training in the army or in jail and indeed in school that Michel Foucault describes is an extended application of these early regulations on the body,[21] especially on touch, of which, we should remember, the form and content is *contact*.

At this point, touch also begins to be a mere preliminary for other actions, especially in fingers and palms: scratching, holding, gripping, hitting, pulling, pushing, rubbing, and so on. Once in this realm of actions, one almost forgets touch; it gets pushed inside the core of these actions; it becomes meaningful, which is to say it loses its primary literalness. From this time onward, we live in a world of meaningful touch, touch always semantically loaded. The relegation of touch to the background of various physical actions also reflects the genesis of an *intention*, directionality of consciousness, purpose transcendent to the act of touching. This generates a third category of meaning, counterposed to the categories of good and bad touch. This is the category of "instrumental" touch.

All these tremendous and serious events in the bildungsroman are accompanied by an allegory of cognition: Clearly touch has an epistemic role. Through the sense of touch we learn of the hardness and softness of things, which is a matter of resistance, and we also learn of relative heat and cold. In the allegory, these primary pieces of knowledge function as metaphors of love, to be slowly and systematically planted onto being.

Eventually, these serve as markers of character. Through this tropology, epistemological categories become ontological.

Touching oneself, we have seen, gives an ontico-ontological solidity to one's selfhood; touching others introduces sociality into the topos of knowing through touch. The genesis and consolidation of purposive action, of motion, of an intention, accompanies this ontico-ontological consolidation. It is important to remember that for this intention to arise, one must already have acquired considerable motor control over one's body. Once the intention is in place, an attitude of the body develops, and at the allegorical level, a will develops and is consolidated. Touch and other senses now serve merely as instruments that this will and intention utilize. What we have called an attitude of the body, its bearing, its directionality—its *disposition*—is described by Merleau-Ponty in his *Phenomenology of Perception*. The disposition of the body is again accompanied by a disposition of consciousness, an intentionality in a loose Husserlian phenomenological sense. (It should be remarked that those people in whom this development is less coordinated tend to mis-cognize the limits of their bodies: These are the people who do not cognize that they are sitting too close to someone else or whose flailing arms hit objects or other persons; they are the ones who step on another's toes without realizing or caring to realize. It is interesting that in society, such acts are quite often misunderstood as aggression and, equally often, generate a counteraggressive response.)

At this point in its development, the self is poised to enter the realm of the social in a new way, with a new understanding of society and sociality as such. However, before we can begin to describe this poised self, we need to describe the development of the sense of space and time, as determined by touch.

THE GENESIS OF CHRONOTOPES

The bundle of synaesthetic experience learns to coordinate and thematize places into organs housed in a body, and in this process, also begins to experience spatiality. It learns that it is a body extended in space. It seems that increasing motor control over its body, especially its arms and legs, is the basis for the experience of space. That arms and legs are extensible means that can touch things and know them as distant objects through the sense of touch. As is well known, in the neonate stage, the visual sense is less coordinated: Anything that fills the visual field is believed to be close by. Enlargement and reduction in size in the visual field is mapped onto spatial distance only later, after the sense of distance and extension is understood. All kinds of optical illusions, including magic, play on relative size in the visual field (and its association with distance), but the sense of touch, especially the sense of touch in extended arms and legs, cannot be deceived as easily and is a surer measure of distance and closeness.

As always, the sense of time is the most difficult to understand, and our opinion is that our sense of time is determined by a physical principle that relates to space rather than to time. Average experience of time as a sequence of phenomena is generated by the absolute physical principle that two objects cannot occupy the same space. They can do so only by sequencing themselves one after the other. Mother's face must disappear from the visual field, and become too distant to touch, before father's face can occupy the visual field and be within touching distance. The experience of space fundamentally determines the experience of time as a sequence. This is augmented by the experience of the reflective consciousness that the process of extending arms in order to touch something is a sequence. We believe that most human experience of time is of this kind. Within the tradition of philosophical discussions of time, one observes the dominant Aristotelian tradition of understanding time through motion, as change from one place to another. Other conceptions of time, which are not vulgar conceptions of time, fall either within what we will soon describe as the chronotope of desire, or some other chronotopes.

The experience of the mutual dependence of space and time, which we have tried to capture in the word *chronotope* (borrowing it from Mikhail Bakhtin's discussion of forms of time in *The Dialogic Imagination*),[22] generates the experience of the body as a spatially coordinated unity of different organs and surfaces, and generates a disposition of the body. Once the body is coordinated and organized in the manner described earlier, it becomes possible to coordinate objects within an abstract experience of distance and space. Within the boundaries of our bildungsroman, every experience is loaded with epistemological and emotional values. Distance and consequent helplessness are to be experienced in the separation from the mother, in the realization of one's body as something other than the mother. One of the earliest things that the neonate learns is to manipulate distance, to bring closer that which is distant, outside its visual and sensory field by emitting high-pitched sounds, and possibly wriggling about at the same time. (This should also reveal that the sense of hearing plays an important role in the development of our sense of space, but in the end, this is only in preparation for the intimacy of lips and nourishment and love.) At this stage, notions of pain and pleasure are mapped onto distance and closeness, especially because closeness often is followed by intensely pleasurable envelopment.

It is not difficult to work out the various stages through which this strand of the bildungsroman will pass. We move to that point of poise where the earlier section ended.

Space manifests itself, phenomenally, as touch. Spatial distance is experienced in something located beyond one's reach. As indicated earlier, one's sense of territory is dependent on what one can touch, first by extending one's arms and legs and so on, and second, by moving as little as possible. Something that we can touch only by moving a great distance in space probably is not ours. That touch and possession are related to each other is clearly visible when one hugs one's friend in the presence of her child, only to have the child rush up to mother and push one away and establish its rights of touching, which it does not wish to share. It is also to be observed that repetition plays a fundamental role in possession. That which we touch repeatedly becomes our possession, imparting a synecdochic bit of our self to that object or person. In many cases, this synecdochic object might be a fetishized object, in which case it can be something that is even outside us but "is more us than ourselves," as psychoanalysis has it in Žižek's reading of Lacan.[23] This is evident throughout the bildungsroman and beyond: toys, clothes, bags, pens, wallets, lovers, personal, and private property. A whole world of objects that are basic metonymies of ourselves surrounds us. We are enveloped by our possessions.

At the social level, groups are constituted by regulations on which persons we can or cannot touch. The beginnings of caste, at the level of the horizontal distribution of regulations on touching, can be located here. In the very institution of the horizontal distribution of rights of touching one can see both the origins of caste consciousness, and caste hierarchies. The upper-caste person would rather not be touched, even metonymically, through the objects that it possesses, or through the metonymic touch of the dalit's shadow. The horizontal/vertical distribution of rights of touching generate norms of commensality and connubiality. The maintenance of spatial distance among social groups is now transformed into hierarchical social distance and distinction. This would allow us to state that caste is, to a large extent, a matter of metonymic organization of objects and persons, if we see its genesis from the genesis of the body, of chronotopes, and of the self and its possession of territory.

The primordial chronotope is that of envelopment, in which time stands still as love envelops us, touching us all over. In actuality, one is not touched on the complete and extravagant surface of the body, but the agency of touch—arms and torso—surround us with overwhelming love. Envelopment is an excruciating, blinding white synaesthesia, where all color distinctions disappear. This chronotope will find various manifestations physical and mental: home, security, warmth, love, trust, and so on. We are speaking here of this chronotope as it emanates love. If it emanates hatred or distrust, we have the Gothic castle, the claustrophobic low ceiling, anxiety, and its existentialist version, *Angst*. This is a deeply spatial experience, one in which time more or less comes to a standstill. All notions of timeless joy or timeless pain are derived from this chronotope of envelopment.

The second, less primordial, chronotope is that of motion. Behind Bakhtin's literary chronotope of the road, or the journey, there is hidden this chronotope of motion. This chronotope is also the chronotope of desire, which is fundamentally dependent on the experience of distance and closeness, of far and near. In this chronotope, time is experienced as memory and anticipation. As is evident in these words themselves, several conceptions of time that are not vulgar conceptions of time are dependent on this chronotope. Those conceptions that emphasize the role of memory in the experience of time (voluntary and involuntary memory, as in Freud, Bergson, and Benjamin), and those that emphasize memory and anticipation and futurity in general (as in Husserl and Heidegger and, to an extent, in Lacan and Derrida) are made possible in this chronotope. If the vulgar conception of time is derived by giving a mechanical scientific interpretation to motion and extension, these conceptions of time are derived from the chronotope of desire.

This chronotope quite often makes us understand some other chronotopes as chronotopes of envelopment. Thus, the experience of belonging to a family is often seen in terms of envelopment. By extension, the sense of belonging to a social group—a matter of caste consciousness—is most often seen in terms of envelopment. This socialized chronotope of envelopment generates rituals that further reinforce the sense of belonging and unity. The sense of belonging, we must remember, is a two-way sense. When we belong to some group, to a large extent, the group's physical and mental possessions belong to us. The group's territory is my territory. Thus, even if some other member of the group does not mind violation of his or her territory, we can "help" them protect their own territory with humble missionary zeal or violent fundamentalist zeal. As is frequently to be seen, these matters of doing are transposed, through a tropology that is yet to be investigated, into matters of being. Thus, the question of religious and caste consciousness is portrayed as the question of "Hindu identity." The primordial chronotope of envelopment relates to being, and through a tropology, other realms are transposed into this realm. We have only touched on some aspects of this complex phenomenon.

There are other chronotopes that we will only indicate here. There is the chronotope of lovemaking, with its entangled and topsy-turvy space and time; there is the chronotope of scientific investigation, with its empty space-time; there is the chronotope of religion, which has dualism as its fundamental principle. In the chronotope of religion, space-time is divided into two kinds, worldly and otherworldly. The otherworldly is quickly given positive and negative notation; however, the single most dominant theme of the otherworldly is god, or its equivalents. At the allegorical level (and we might want to wonder whether it is not the chronotope of religion that makes allegory possible in the first place), which is also to say at the philosophical level, we might call this the chronotope of the transcendental signified.[24] We can see quite clearly how this chronotope is partially derived from the chronotope of envelopment.

There is also the chronotope of exchange, and this needs more attention. In the economic encoding of exchange, time is not taken into account. Economics cannot comprehend a delayed exchange; it relegates such exchanges to the realm of gift-economy, leaving that in turn to anthropologists and sociologists.[25] In the economic conception of exchange, there is no temporality; within it delay is incomprehensible. It might seem that futures-markets are capable of conceiving delay; however, the exchange in these markets is instantaneous; only the delivery of goods is delayed. The same is true of use-values. Delayed use-values are incomprehensible within this chronotope: If we buy something in order to consume it, say, fifteen years later, it becomes an investment, or an act of hoarding, not of consumption.

This chronotope has the intensity and immediacy of synaesthesia, but its constituents are the various metonymies of our actual and possible possessions. Objects produced by labor become commodities within this chronotope. We have already indicated the relation between the realm of objects and object-commodities that surround us with their closeness to fetishes and the chronotope of envelopment. It is also clear that this chronotope is linked to what we have called the chronotope of science, with its empty (and therefore pure) concepts of space and time. It becomes clearer now that the various chronotopes form a kind of gigantic multiworld system that our body traverses continuously. We suggest that many of these chronotopes can be seen as allegories of touch.[26]

2

TOUCH—AN A PRIORI APPROACH

THE A PRIORI DIAGRAM OF TOUCH

It is possible to posit a theory of genres of touch, based on the discussions in the preceding chapter. Some features of this theory have already been indicated. The primary categories of genre classification here are the following pairs of concepts:

1. touching oneself / touching others
2. literal touch / figural touch
3. good touch / bad touch

We also suggested another category of touch, that of instrumental touch. For the sake of convenience and elegance we categorize this type of touch under figural touch. Some argument too could be made for doing so. The purely instrumental touch, that transparent touch that all of us experience when holding food, for example, or pulling down the hem of one's dress, is a touch that is meant to do something more than touch food or hem. The touch is turned away from itself, which allows us to categorize it as figural touch. However, this figural touch cannot be given any name from among those available within linguistic tropology. We may tentatively call it the zero-degree touch.

The interrelations among the genres of touch need to be schematized; the structure of the theory represented in a diagrammatic form. It is always possible to write down the genre-theory in plain or moving prose; however, the essence of a theory is expressed nicely in a synoptic, synchronic diagram, especially because we cannot write in the language of touch.

In constructing a diagram of the genres of touch, we follow the interesting diagram constructed by Franz Stanzel in his *Theory of Narrative*,[1] which is that of a circle thrice divided into two parts, so that the diagram actually consists of three circles, each divided into two. The divisions are based on oppositions, and the lines dividing the circle function as a kind of boundary. Until one reaches the boundary, one finds some continuities along the periphery, which is true also for the other half of the circle. Where the two parts meet on the periphery, there is ambiguity regarding the kind of narrative placed there, as to which part of the division it belongs to.

We give below a diagram of touch that is based on a simplified version of the complex diagram devised by Franz Stanzel. It should be noted that in models that use the circle it is often construed as a metaphor of completeness, wholeness, continuity, eternity, cyclicity, envelopment, and so on, but there is another interesting aspect to the circle, and this is that all the points on the circle, that is, the periphery, are equal. In the context of the diagram that we are constructing, this observation is quite important because it is not as metaphor of completeness or envelopment that we use the circle, but precisely for the bland equality it imposes on all the points on the periphery. This is a little like initializing a system before one starts using it. The bland equality deprives the privileged points of their privileges, and equally blandly, deprives the underprivileged of their underprivilege. All values have to be reset to zero initially, for a "revaluation of all values" to be possible. This is an irreducible moment inevitable in all serious re-valuations. This is a moment of the greatest of poverty of meaning and value.

Diagram 1 represents a possibly simpler way of thinking of the relationships among the various categories. The representation itself is not the most significant issue, though it should naturally strive (but always fail) to be an exact representation. It is this simpler version with which we will work. There is some argumentation possible as usual: The fact is, goodness and badness are qualities, and by making them into firm oppositions, we run the risk of making them into substances, which they are not. Also, notions of goodness and badness of touch are more variable than notions of touching oneself, touching others, and literal and figural touch. These categories of touch are quite clearly more substantial. Therefore, it seems wiser to use the latter pairs as oppositions and attach the qualities of goodness and badness to them. This is also a simpler version of Stanzel's diagram. It should be constantly borne in mind that the diagram is only a convenient machine; there

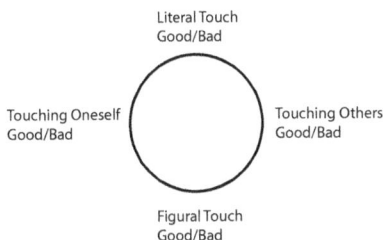

Literal Touch
Good/Bad

Touching Oneself
Good/Bad

Touching Others
Good/Bad

Figural Touch
Good/Bad

DIAGRAM 1

is no plenitude of metaphoricity in it; it is inherently poor in meaning, though marginally richer in order.

THE FIRST OPPOSITION: TOUCHING ONESELF AND TOUCHING OTHERS

We now discuss the three oppositions independently of each other, devoting a section to each, in order to clarify them and observe the manner in which they operate in society. In keeping with earlier procedures, we begin with the opposition that is most obvious and ostensible, that of touching oneself and touching others. Although we follow tradition in beginning with this overworked opposition between self and others, it seems wiser to do so, since the question of values, of good and bad, is rather difficult. It is wiser, since most people know the difference between touching oneself and touching others, or other things. Edith Stein, in her *On the Problem of Empathy*, points out: "I could never say that the stone I hold in my hand is the same distance or only a tiny bit farther from the zero point than the hand itself."[2] Learning this difference and this distance is one of the major processes and achievements of the neonate. In fact, the neonate becomes a baby, and later a child, in the process, as we have already argued.

What is the difference between touching oneself and touching others? An answer to this question is also a partial answer to the question What is the difference between the self and others? (We say that it is a partial answer, because the two questions are not identical.) We have already seen that in relation to touch, the sense of self develops in a specific manner. The sense of the self is fundamentally dependent on the sense of proximity, and it is almost impossible to distinguish the sense of touch from the sense of proximity. That which we can touch easily is proximate, and there are very few things which are as proximate as our body. Explanations, interpretations, or descriptions of the relationship between self and others that are based on the visual sense are fundamentally deficient in this regard. The gaze of a formed or forming subject on the formed or forming object is not a useful model or metaphor for understanding the relationship. The ideas of self and others are generated and disseminated through the experience of touch. As we have already tried to show, the neonate's sense of self consolidates itself through the experience of touching, or failing to touch, others.

In this realm of touch, a primary categorization is that of touching and being touched. The senses of activity and passivity are also related to touch in this sense. Touching and being touched constitute the primary cognitive realm. We believe that it can be observed that touching, fused as it is with some activity, quite often reduces touch to what we have called the instrumental touch, a purposive, pragmatic touch. Being touched, in contrast, has a stronger cognitive aspect and alerts our senses in fundamental ways. The active touch is usually in the service of some other purposive action; the passive touch makes our body sensitive to otherness. But have we not, in thinking with the categories of activity and passivity, already taken up, without examination, two categories fundamental to Western philosophy since Aristotle, categories, moreover, that presuppose the distinction between self and others that are yet to be derived? On the contrary, it seems possible to generate notions of activity and passivity on the basis of the experience of touching and being touched.

One identifiable difference between touching and being touched is that between expectation and surprise. If we undertake some motion of the hand or the lips, or the whole of the body, expecting to experience this or that kind of touch, and the expectation is fulfilled, then we could say the experience is one of touching. Whereas others always surprise us by touching us.[3] The experience of being touched is that of surprise and can often result in a preternatural twitching of muscles. It should be noted that something similar may happen when one has undertaken some motion expecting one kind of touch, and one experiences an "unexpected" kind of touch—picking up a vessel that one believes to be merely warm but which turns out to be scalding hot can serve as an example here. The unexpected and alarmingly sexual touch might serve as another. The point is that the unexpected touch tends to have its origins in something other than one's known and reflective intention; in fact, it has unknown origins.

The epistemological difficulty in knowing one's intention, which psycho-analytical approaches introduce, does not pose any difficulty for this descrip-tion for the very simple reason that though psychoanalysis finds its ultimate basis in drives or "instincts," the body-as-such in its materiality really has no place in psychoanalysis. In it the body is made into a completely psychic en-tity (psychosomatic, to be sure, but it is precisely the point in psychoanalysis that one does not always know where soma and psyche begin and end), a field in which the play of sexual and other energies takes place. It is difficult, within psychoanalytical approaches to think of something being entirely or purely somatic. The body is, in these approaches, a somewhat mysterious point of origin of instincts or drives; this mysteriousness of origin is a Ro-mantic notion (the "id" as the primordial, "inherited" part of the psychic apparatus).[4] To use words that are perhaps unfamiliar within this type of discourse, our attempt here is to speak of low-level programming, at the level of assembly, rather than developed programming languages as such. The body is assembled before a psychology can take hold of it. It is not surprising, therefore, that for Freud, and for Lacan too, the "real" is an "algebraic x."[5] There is another point to be made: The "unconscious" becomes relevant and evident only inasmuch as it disturbs the syntax and semantics of conscious behavior; therefore, it becomes clear that Freudian psychoanalysis privileges conscious behavior (inasmuch as the attempt is to understand and "correct/cure" uncommon conscious behavior). If there were no slips of the tongue, jokes, or other disturbances of the syntax of conscious life (semantic distur-bances are deduced from syntactic disturbances)—in short, if there were no symptoms at all—the unconscious would remain undiscovered. It is possible to believe and therefore state that it does not matter whether the motion that we undertake is undertaken out of conscious or unconscious intention; the point is that motion is undertaken, whatever the type of intention.

Once action has been undertaken, the conscious or the unconscious (it does not matter which), expects certain kinds of touch, and if this expecta-tion is not fulfilled, then there might be a knee-jerk reaction, and the touch experienced is likely to be experienced as hostile. There is a primordial "re-sponse" here, which can be found in all animals, and in some plants as well (the touch-me-not is just one poetic example; the carnivorous plants like the Venus fly-trap are another). It is to be observed that the matter of "response" is not just a matter of the language-game that we play or the discourse within which we are trying to make sense of touch. It's not a matter of making statements about touch or treating touch as an object of our discursive endeavors. It is not absolutely necessary to believe that response is possible only to animate beings. Chemicals, subatomic particles, electrical charges, and memory alloys—most existing things respond in some way or other. Primates might have variations in their responses (which makes them a more complex, and therefore an unstable and unpredictable, system), which variation of response requires, given the limited and inherited vocabulary that we as human beings possess, the appellation of animate. Inanimate things tend to respond without any variation or unpredictability, and the relation between the self and others has to be located in this area of unpredictability, surprise, and wonder. This invariance of response in the material world is the prime object of the physical sciences. As is well known, as variability increases, there is a softening of the sciences. It should be remembered, however, that there is, even within the body a level of invariance. The medical sciences work in this area.

As already stated, an unexpected touch is likely to have a source other than an intention known to us consciously or unconsciously (and this could be within me or outside me). The actual neurographic patterns involved in knowing one's intention are not really important, since what is important for our discussion is the matter of expectation, of what Heidegger and, following him, Gadamer call "fore-understanding."[6] We believe that the complex business of repetition too can be encoded under the heading of expectation, since without expectation and foreknowledge or preunderstanding or prejudice or forehaving there cannot be any repetition. Prima facie, it looks as if repetition is a matter of memory rather than expectation. No doubt, to a certain extent this is true. However, there cannot be a repetition unless something is expected and the expectation is fulfilled, with but a variation. The variation is significant, since if there is no variation, repetition becomes identity.

One of the most remarkable events in history is that invariance in response has served as the final paradigm of knowledge: Laws of nature are representations of this invariance of how inanimate things behave. In this realm, it seems, there is no repetition; there is only identity of response, which is then lifted up to the level of a principle, or a principle of explanation, such as gravity, principles of dynamics (solid or fluid), and so on.

We have already indicated that animate response tends to be variable, and it is here that expectation and its partial fulfillment might be located.

We might remind ourselves of the various games of unexpected touch that children play, games in which one is to steal up to another person and touch him or her stealthily, so that he or she experiences unexpected touch, and thus is, caught or is said to be out. There are various versions of this game, some employing visual cues rather than touch.

When one's hand reaches out to the bolt to slide it in order to open the door, one knows the kind of touch that one will experience. This knowledge is actually nothing more than an expectation of a certain kind of touch, which expectation, as always, has its basis in memory. If, however, we are opening the door in the dark, and instead of the bolt with its expected texture, solidity, and relative coolness, one touches a slippery texture that is softer than expected and much cooler, and a hissing noise seems to come from close to one's fingers, then one is likely to scream, realizing that some snake- or lizard-like creature is sitting on the bolt. This is an example of what might happen when the expectation is shocked, negated. One should, however, also observe what happens when there is no memory to guide our expectation. One might try touching an object that one has never touched before, and one will experience otherness in a fundamental and wonderful sense, and this sense of otherness and wonder is quite beyond the values that the touched object may have for other people. One might observe children touching kittens for the first time in their life, the hypothetical alien encounter, the first sexual kiss, or similar experiences.

This slender evidence can be multiplied with a little imagination and observation. It is in this interaction between memory and expectation that the temporality of touch may be located. This allows us to see that in this realm, otherness is as much a temporal experience as it is spatial-differential. In the unexpected touch by others, we experience a quickening of time and heart; in the expected touch, depending on the intensity associated with it, we experience a slowing of time. This lengthening and shortening of time is most clearly experienced in sexual touch. Touching oneself, because of its almost subliminal nature (which is different from the Freudian unconscious), tends to be without any substantial temporality—the surfaces of our own body that we touch tend to be more or less the same throughout. No doubt memory imposes a pattern on the experience which quite often prevents us from touching oneself anew. One might go so far as to say that one can experience something new while touching oneself only in the pubescent period of growth, and in the condition of bodily disease. Touching one's wounds thus forms an opposition with the pubescent experience of touching one's growing breasts or hardening penis. These experiences, because of their unexpectedness, render one's body into a curious object, an object of unfamiliar cognitive values. The experience of touching oneself, otherwise completely familiar, is restored to us at such moments by the process that literary critics call de-familiarization,[7] though in this case, the de-familiarization is caused not by a change in point of view but by an objectively different sensation.

It is worth considering whether it is possible to divide touching oneself into styles (we use this word to indicate the difference between this lower-level categorization and the kinds of touch). Touch, we have already seen, is one of the fundamental senses that give us our sense of self in the first place, and it is not unexpected that we should hardly ever really notice when one touches oneself. Here too, knowing that one is touching and not knowing it form a major division. Let us tentatively say that this division is between unreflective and reflective touch, since knowing is always enveloped in reflection. Of the reflective touching oneself, we could say there are a finite number of styles. Let us enumerate some of them. Since attaining childhood, locomotor independence, and relative freedom to look after one's own body, one is made to touch oneself in order to clean oneself. Washing oneself forms one major chunk of our touching ourselves throughout life. This touch that takes care of the body, which renders objective the body itself, which examines by touching various nooks and crannies and crevices of the body (many of them never to be seen with one's own eyes without some reflector/ instrument), itself needs further subdivisions, since not all touching oneself to look after oneself is absolutely necessary, nor do all these touches have the same value or significance.

One might want to consider whether bathing—that great purificatory action in the Indian tradition especially so in the brahmanic—is absolutely necessary. (This helps us understand why washing is supposed to purify oneself, especially if one has been contaminated by a dalit's touch: A knowing touching oneself is the exact structural opposite of the unexpected touch by others. Subsequently, metonymies develop, so that an object touched by a dalit is washed to re-purify it.)[8]

We must distinguish between norms of hygiene and necessities of health that are essential for continued survival. Here we use the word *hygiene* as distinct from health. It is obvious to anybody who has not bathed for a day that bathing is not absolutely necessary. This act of health care turns out to be more hygienic than expected. From the point of view of health as such, hygiene looks more like cosmetics. In short, we touch ourselves more in terms of hygiene than health. This becomes clear when we are ill or wounded. With what great care and sensitivity does one touch one's wound. If one cleaned oneself with the same intensity, one would probably never ever need to bathe again! Thus, we have three styles of touching oneself in the general category of looking after oneself: the health-related, the hygiene-related, and the cosmetic-related. All three originate in society and are given to us by others. Health comes not from within one, but from someone else, just as hygiene and cosmetics do, a little more obviously. Both share the same relationship: These three styles are touch for-others, and not for-oneself.

Another style of touching oneself is that of sexual pleasure. Masturbation serves as the best example of touching oneself sexually, though there are variations on this theme. It is to be noted that this kind of touching oneself is accompanied by an intense solitude (even if it is done in relatively unisolated surroundings, or under a camera, for that matter). The dominant tone in this style is that of guilt, again socially given to us. A great many people find it extremely difficult to un-guilt themselves in this context, though in other areas of social action they may believe themselves to be, or be, above the social norms, or even beyond good and evil. This harmless pleasure that does not harm others or oneself has some other implications. It is incontrovertible that the genitals are loaded parts of the body. They get encoded as parts, rather than as mere surfaces on the body early on in the transition from babyhood to childhood. These parts, incidentally, do not seem to be entirely controlled by one's will, or intention. Quite a lot of especially male insecurity could be said to spring from the realization that the penis has its own sweet will and may not always obey commands. Therefore, touching these parts is always the most intense kind of touching oneself for pleasure. It is important to realize that this kind of touching for pleasure is entirely metaphorical. We will discuss the metaphorical-allegorical nature later on. The notion of uncontrollability seems to be a legitimate part of touching oneself for pleasure.

What should serve, or be made to serve, as the best example of touching others for pleasure? It is not so unremarkable that here too sexual touch is the most likely and suitable. We need to make crucial decisions here, since it is possible to bring in age-old questions into our discussion: Is not the mother's touch as significant and intense as the sexual touch? Is not the altruistic touch—picking up someone who has fallen down or is about to do so, and so on—just as significant as the sexual touch? How does one, meaning anyone really, living in the times that he or she does, even begin to wonder about which touch is most important for him or her? It is here that we take a line of thought that goes against the current perception of the importance of sexuality.

Although in the final analysis the sexual touch is something that serves either reproduction or pleasure, it can be pointed out that the altruistic touch is essential for survival itself. No doubt there are complex issues involved in the notion of survival (and we might mention Richard Dawkins's *The Selfish Gene*)[9] for a particular organism with its different and yet similar genetics to come into play in the game of survival; it is essential that this organism reach the stage of sexual maturation, and it is the altruistic touch that is the most significant from this point of view. This type of touch is not always the same as the literal parental touch. One only has to go through John Boswell's *The Kindness of Strangers*[10] to see that the altruistic touch, literal or figural, is possibly fundamental for the adopted children and those who adopted them.

Let us, however, return to our own concerns. We mentioned John Boswell's remarkable work here only to remember that the caring or the altruistic touch is not always necessarily equal in meaning or always reducible to the parents' or lovers' touch. This allows us also to move into the realm of sociality proper, away from sexual concerns, which, though intense, tend to center on one's own individual experiences. Possibly it is somewhere between the sexual and the altruistic that we need to locate touching others. Taking altruistic touch to be fundamental and sexual touch fundamental reveals that social inequality must express itself in terms of these two: Between the brahman and the dalit an altruistic touch is impossible; so is sexual touch. Thus, a violation of this impossibility must negate these two types. This is, as we see in society, precisely what happens when both the impossible altruistic and the impossible sexual touch are actually practiced: There is violence among men, and there is the rape, or sexual exploitation, of dalit women. For us, living in the educated world of the twentieth and twenty-first centuries, sexual touch is almost always accompanied by love, however momentary it might be. Rape and sexual exploitation are its opposites.

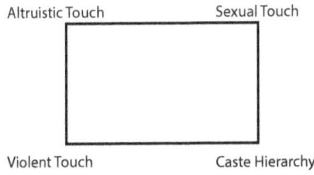

Altruistic Touch | Sexual Touch

Violent Touch | Caste Hierarchy

DIAGRAM 2

Thus, touching others could be said to be built up of the possibility and/or impossibility of the altruistic/caring touch and the sexual touch. The caste hierarchy represents the opposite pole of the altruistic touch, whereas the rapine or the exploitative touch represents the opposite pole of the sexual touch. This can be represented by a diagram. Please see Diagram 2.

We have already discussed the relationship between touching and being touched; we have located being touched as one kind of experience of otherness, and touching others as another kind of experience of otherness. It is always possible to see that the experience of otherness in touching is always in danger of reducing otherness to a particular example; moreover, it is stamping that other with an objective quality. This stamping is always fraught with metonymies, and as we have already argued, it is fundamentally related to territory and selfhood and possession. Being touched, in contrast, makes one realize the reciprocity of this objective stamping, and thus is a more fundamental experience of otherness. The argument here is that both activity and passivity give us an experience of otherness but of different kinds, and one of these, being touched, is a more fundamental experience than the other, touching. It is to be noted that we state this not only for any ethical values such a conception might have—the "totally other is unknowable but is to be respected" type of value—but for the comprehensiveness that this formula seems to possess. In fact, it can be seen quite clearly that those philosophies that value "the other," or in a more precise formulation, "otherness" more than the self are likely to be altruistic in the sense in which we have used the word.

At a later stage, the social and sociological implications of the diagram will be given; at this point, the diagram as it stands without commentary suffices to give some idea of the dichotomies involved in touching others.

THE SECOND OPPOSITION:
LITERAL AND FIGURAL TOUCH

Even more than the first opposition, the opposition between literal and figural touch seems obvious; and in fact, it provides the first insight into the questions of caste and untouchability. We have already described some basic features of this distinction in chapter 1. Let us remember that the literal touch has the structure of an infolding, an invagination, and therefore is difficult to see sometimes because of its sheer visibility. Literal touch is like the thing that stares us in the face, but which we cannot see. It is always possible to confuse this excessive visibility with blind spots, which could further be attributed some function of negativity built into the structure. This confusion must be carefully avoided, and literal touch seen for what it is.

How does one ever get to know that there is literal touch and that there is figural touch? We get to know literal touch in two ways: through sensation, and through the difference it makes when compared to figural touch. Both these provide legitimate access to the literality of touch. It would not do, again, to think that it is only the difference from figural meanings that establish literal meanings to touch. There is, after all, the touch of the blade of grass or mother's powdered cheek; there is the smarting pain as well, caused by a variety of things touching the skin on various parts of the body, of which the slap on the cheek is just one quick example.

It is here that we must be careful in not assigning metaphoricity or figurality of any kind to this intense experience of touch. In the realm of touch, we will have to argue, literal touch is as intense as figural touch, if not more: It is easier to allocate intensities to figural touch. In fact, the intense literal touch—sexual or otherwise—is set in the foil of literal touch of rather low intensity. It is this low-intensity literal touch that is the easiest to forget, or neglect; therefore, we need to pay rather close attention to it.

It is not really surprising that our survival depends on this low-intensity touch, which has been relegated to the rather mundane task of being constantly alert to dangers. The pin on the floor, the broken springs hidden under the cushioned seat on a chair, the innocent-looking brush of fingers on the exposed midriff, between blouse and sari, the unexpectedly aggressive grip on the neck—all these cognitive and survival-related sensations are basically of the low-intensity type. These are active even in sleep: In the cold, if the sheet slips off, one wakes up, partly to cover the exposed bottom, or again, if a centipede crawls over the blanket, one is woken up by even those minuscule vibrations (though naturally the blanket must be thin enough to carry such vibrations). It is these that alert the organism to possible dangers. From here, the metaphorical values seem a luxury which only the upper castes can afford.

Metaphysically speaking, it is this literal, low-level type of touch that generates notions of empiricality and reality. Concentrating on this stratum allows us to see that the notions of empiricity and reality themselves might be generated in a variety of ways.

Let us take a partial look at two texts to illustrate this point. The first is from Godse Bhatji's *Maza Pravas*.[11] In 1857, a brahman from Varsai near Pune traveled north. On the way up, the traveling party met two soldiers who had deserted the Bengal Native Army, just before the 1857 "mutiny" against the British, and they cautioned the travelers that there was trouble brewing. They told the following tale, which goes something like this in summary. A brahman soldier went to the river to drink water, and so did a *chambhar* (cobbler, often lower caste) soldier. The chambhar asked for the brahman's jug to drink water from. The brahman refused (presumably because the chambhar's touch would contaminate the jug). The chambhar is reported to have said to the brahman, "Don't be so proud of caste; all that is empty show. The bullets that you put in your mouth are covered in cow fat and pig fat, which cow fat and pig fat I have handled." The brahman got angry; they came to blows. Other soldiers gathered, realized that the very *dharma* was threatened with contamination, and the first sparks of the conflagration of 1857 were lighted.

What is germane to our purposes here is the sentence, reported by the deserting soldiers, which the brahman soldier uses to refuse the jug: "I cannot bring myself to give it to you." The brahman soldier (here, B) is behaving according to a norm that is for him, and according to him for others too, transcendental, and therefore the curiously passive form of the sentence. B has been living his life according to those principles. The chambhar soldier (here, C), however, does not take recourse to any abstract notion or transcendental norm to rebut B. C's reaction is based on his empirical, and therefore what is for him undeniable, knowledge (so different from the knowledge that B possesses) that the bullets are covered with cow fat and pig fat. Within his universe of discourse, his knowledge is undeniable. Presumably, for B too, within his universe of discourse, his knowledge is undeniable—only to a certain degree, however. B does not disbelieve or disprove C. At the ground level, it seems, the type of knowledge that C has is unbeatable. This is also the purport of some of the questions that the Carvaka tradition asks the Vedic tradition, which believes that things sacrificed in the sacred fire go to heaven. If the thing put in the fire (sacrificed) goes to heaven, why does not the sacrificer put his father into the fire?[12]

The other instance is of a song, recently recovered in audio compact cassette format. The tape itself, published by the Centre of Indian Trade Unions (CITU),[13] a leftist union, says of this song that it is originally a Telugu song. The words themselves—in Marathi translation—are at least as interesting as the musicality of the song (its type and quality). The song is formed as a series of questions; the song's rhetoric puts caste distinctions under question: "We broke our backs tilling your land, and you ate of the grain; how come you did not ask then what our caste was? We collected flowers for you; you put them in front of the god; how come god was not contaminated?" or, "We pulled the dead cattle, and you used the footwear (made from leather); how come you did not ask then what our caste was?"

As always, empiricism of this kind speaks for itself, especially in the contexts of real social injustice. Political rhetoric always aspires to the power of such questions: In short, then, this type of empiricism is political through and through. These two examples reveal, we hope, that empiricism itself can be made to relate to ethics.

There is another point to be made about this kind of empiricism and, by extension, materiality: All forms of production involve touch. As indicated in chapter 1, we are looking for a notion of materiality that is at least one level below the traditional notion of materiality in the Marxist, or even the non-Marxist, aggressive consumeristic sense of the theme of *carpe diem*, of seizing days and pleasures, such as our society affords us. This somewhat trivial observation allows us to see that an organization of touch and its regulation must determine forms of production, which, in a sense that should satisfy traditional Marxists, subsequently organizes societies around the means and relations of production and their control. For the sake of clarity, let us assert again: All forms of production involve touch. Marx observes in the very second paragraph of *Capital* that to discover the various uses of things is the work of history. A different use of a thing can be found only in manipulating it differently, which can be found only by touching it, twisting it differently. (It should be possible to see that Claude Lévi-Strauss's notion of *bricolage* is potentially present here in Marx's sentence.)[14] It is no doubt quite easy to imagine a new use of a thing, or a concept, in which case it might not be correct to insist on the manipulation by touch. However, it should be equally easy to see that apart from the realm of some "pure" philosophy, most forms of production are dependent, in the final instance, on touching an object. It is of course entirely possible that one manipulates objects that in turn manipulate second-order objects and so on. The analogy with "the production of the means of production" is valid here.

It is important to appreciate this. The sensation of touching is as indubitable as the consciousness of being conscious. Just as my awareness of myself thinking is indubitable, so are sensations indubitable, and this includes dream sensations, while they last. A subsequent modification or reallocation of certain sensations as dream sensations, or as fictional sensations does not subtract anything from the reality of sensations as sensations. The business of doubting the reality of sensations we have foreclosed in the first section. It is not our intention to discuss the business in detail here, and as has been our practice, we have briefly indicated some of the ideas involved.

Production, which is more germane to the issue here, involves touch, and it follows that regulations on touch also regulate production. Only certain kinds of production are possible given a certain organization and regulation of touch. If only a certain group of people is "permitted" to touch certain kinds of objects, only members of that group will be able to use the objects, according to convention, or to change convention and make history. A transformation of such regulations is a transformation that digs deep into the very structure of society, working at a possibility of the transformation of that structure itself.

In short, a caste system itself can always be thought of as a structure of relations of production (understood here materially as who is permitted to touch which kinds of tools and objects, etc.) and an entailed or dependent ideology, and there is nothing wrong in this conception.[15] As can quickly be seen, however, our wholehearted acceptance of this statement leads to certain anomalies. If caste is thus reducible to class through the notion of ideology, then the following question arises: Can there be a class ideology that does not acknowledge the existence of classes? (Since caste is class, and caste is disacknowledged, if the two are the same, the disacknowledgment of one is the disacknowledgment of the other.) Or a class ideology that does not even think in terms of classes? We seem to be faced with a contradiction, which could be a contradiction either (a) in the position which sees things in terms of caste, or (b) within the very notion of ideology itself, which perceives and conceives ideology as something premised on, (relatively) dependent on, the material, economic relations of production, and to some degree or other, counterposes the notion of ideology to that of knowledge or science. It is clear enough that in Indian society at least, caste-consciousness (which is socially produced but attributed to birth, as we have tried to show in chapter 1) is, in sociohistorical terms, more common than class-consciousness.

Caste might allow non-Indian thinkers, especially Western thinkers, to posit Indian society as that nineteenth-century, despotic, localized, almost monadic village society, by which presumably it is feudal society that is meant.[16] However, the possibility of Indian "misapprehension" of class-relations as caste-relations needs some description and explanation. It is here that the anomaly is most clear: If caste-consciousness is merely ideology, we should be able to explain why this particular structure of relations of production should generate this particular caste-consciousness. The notion of allegory does not seem to be very useful to such an explanation, since the *allos* of the allegory, the otherwise of the "speaking otherwise," seems to be rather too different in this case. Lacking this description, one might posit a radical difference between the two.

Let us think again of C, the chambhar who had said to the brahman, in effect, "Why do you make so much of your purity? These are all false ideas. The bullets that you bite with your teeth have cow fat and pig fat on them." C seems to be saying, "I *know*, because I extract the fat from these animals" (which, presumably, are killed and/or dead when he does so). It is the confidence in the knowledge that we wish to concentrate on. There is little or no anxiety or passivity in this knowledge, and that contrasts well with the brahman's passive sentence, "I cannot bring myself to give it to you." C's knowledge, it seems, is far less dubitable than B's. It should be possible, in principle, to make this knowledge serve as the plinth on which to build whatever edifice one wants. Such exercises are not uncommon. Dalit philosophy, of which Kancha Illaiah's *Why I Am Not a Hindu*[17] is a less than successful example, could be easily imagined to be filled with such celebrations and/or glorifications of dalit knowledge, its modes, and its powers, if only something existed which could be called dalit philosophy. Let us, with much less ambition, revert to C.

It is clear enough that the indubitable piece of knowledge is put to rhetorical use, directing it at social relations, and thus at a claim to some kind of power, or, to be more precise, to a resistance to it. The falsity of the idea of brahman "purity" is not self-evident to C, but forms a piece of knowledge in which there are at least three atomic units involved: (a) the knowledge that cow fat and pig fat are used to jacket the new bullets, (b) the knowledge that brahmans are forbidden to eat any meat product, especially that of cows, and (c) that the brahman soldiers have been putting these things into their mouths every time they bit away the bullet wrappings. Allow us to speculate further. It is a remarkable fact that this piece of knowledge, which must have been available to C from earlier on, is deployed only when he himself is deprived of something because of the brahman notions of purity. It is only then that this piece of knowledge is brought out and activated. Until then, C is presumably silent on the subject. The complexity of his presumed silence is something that we need not speculate over at this moment. What is equally remarkable is that the knowledge is revealed and deployed at this particular moment. This is what makes it overtly and clearly rhetorical. We are trying to establish that this rhetoric that C employs is supported by his knowledge, which is generated by regulations on what brahmans can and cannot touch: Had they been permitted to touch dead animals, the relations of production would have been different. This possibly demonstrates that the regulations on touch in society make for particular relations of production. It is to be remembered here that at least in the times that we are speaking of, caste is something one inherits even before the profession.

Until now we have discussed the issue of caste and regimes of touch without thinking about the question of religion at all. We believe that religion becomes particularly relevant for attempts to understand the distinction between literal and figural touch, as also literalness and figurality in general. Within the realm of religion and faith, that which is literal for one can be figural for another, as is seen in the debates around the literal and/or figural meanings of the sacraments in Christianity. Faith transforms figural meanings into literal meanings. What is literal for one generation might become metaphorical for the next, as indeed has happened in the case of religion, which in the modern period is seen as nonliteral, figural, and in the case of modern science, which presumably is thought to deal with the literal meanings of natural objects and phenomena. It is from this opening that we would enter into the question of religion.

It becomes clear that the opposition between religion and science, fostered by both the discourses, is based on a disagreement about which part of human experience is literally true and which only metaphorically, or allegorically, true. We have already seen in our first discussion of figural meanings that religion, with its dualism of the this-worldly and the otherworldly, fosters an allegorical interpretation of the world and phenomena within it. In contrast, scientific thought, with its emphasis on literal meanings of life, nature, and human existence, seeks to establish a literal meaning to its utterances. It is worth noting that both these discourses make similar claims toward a universality, a catholicity, so to say, of their own truths. These claims are made viable only by a detour into strong or weak forms of idealism. Through its idealism, science loses touch with materialism, and divides itself into two extreme poles: the blank universality of mathematical statements and the semantically poor, mechanical materialism which can speak only of chemical reactions, the HPA axis (the hypothalamus-pituitary-adrenaline axis) in human behavior of what we have earlier synecdochically indicated by the words *blood, plasma, bones*.

An interesting aspect of the antithesis between literal and figural meanings is that this is a social matter. When a certain group of people asserts the literal truth of the sacrament, or of idol worship, it is their own shared sociality that is asserted and celebrated. A community of believers is formed. It is the sharing that generates the truth-value of the literal meaning. It is not without reason that "literal truth" is a common usage in language. This is very clearly evidenced in the various sectarian differences within the various particular religions. The bread, for some, is literally transformed into the body of Jesus; for some, it is a figural body (though of course no less powerful). For some the vampire is just a fictional convention; for some, it is the truth itself. For some, god is real and literal; for some it is a fictional, if not an ideological, construct. In both assertions, a certain kind of community of humans is operative. Outside this community, however, things become metaphorical, and might even be projected as metaphorical (as in early history of Christianity, the charges of cannibalism were refuted by stating that the business of eating the body of Christ was only metaphorical, whereas in those days it was the mark of faith that it be understood as literal).[18] Socially speaking, then, the difference between literal and figural meaning marks the boundaries of communities. (Incidentally, it should be possible to understand the various arts, especially the linguistic ones, in terms of this formation of community around the literal/figural distinction.)

Ritual (inasmuch as it is an integral part of religion) reinforces the distinction between figural and literal by congregating the members of the community. In ritual, the very process of forming communities around values that are constantly in danger of being relegated to figural meanings is reiterated (hence their frequency), and these endangered meanings are literalized. The first, or the last, human on earth needs no ritual, for the simple reason that there is no community.

The degree to which a particular community is ritualistic and the values around which rituals are built would give us another measure of the flexibility of values. All communities have rituals of this or that kind. That which is literal for modern societies of acquisition is also a value, and that value needs to be reinforced. Since there are no communities without ritual, it is necessary to understand which values are thought of as literal and which figural, and given our framework, such allocations should also yield the primary genres of societies. It becomes clear now that what would be flexible in the first place is the allocation of meanings to the categories of literal and figural, and not the categories themselves. It would be impossible to do without them, tightly bound as they are with notions of truth and reality.

Genres of society can be derived, for example, on the basis of which aspect of human experience is elevated to the status of the real or the literal meaning, and the intellectual justification for this elevation. As is well known, in Platonic philosophy, the status of sensory experience is, in a certain sense, subreal, lower than the reality of Pure Forms; whereas in our modern world, the status of sensory experience is that of the real, and the status of ideas and such like is that of ideality which is given a supra-real or subreal status depending on the philosophical framework used for justifying the ways of reality to ourselves. We can only indicate the possibilities here; let us now return to our main concern.

In a society in which sensory experience is relegated to subreal or supra-real realms, literal touch would have, in principle, no meaning or zero meaning. It is to be noted that no such societies can exist in practice. Sensory experience might be relegated to the realms of bad figural (subreal), or good figural (supra-real) but cannot ever really be completely devoid of meaning.

Literal touch, therefore, cannot ever be devoid of any reality-value without violating the principle of noncontradiction. The sensory experience of touch provides us with that inflexibility that might be made into the basis of all changeability and variation.

We now discuss figural touch and its styles and tones. The analogy with linguistic tropology is not exact; the two sets cannot be mapped onto each other point by point. We have used the dalit's shadow as the basic metonymy for figural touch, indicating in passing metaphorical and allegorical touch. Figural touch is that in which literal touch is turned away from itself, toward something else. Clearly, that touch which is in the service of some other motive, intention, purpose, or action is figural. As mentioned earlier, it would not be possible here to list all the figures of touch. We concentrate instead on metaphor, metonymy, and allegory. We use the word *metaphor* not in its general sense (anything that is a carrying across), but in the restricted tropological sense of a figure of analogy.

That touch is metaphorical in which touching one thing is analogical to touching some other thing. This enables us to see that this is an area of substitutes as well: The simplest example we can think of is that of the baby sucking its thumb. There can be found several such examples of touching a substitute object, especially in the realm of sexuality. Thus, in societies with rigid sexual norms, holding someone else's hand amounts to having sex with him or her. (This is a clear case of metaphorical touch but can easily be mistaken for metonymy. We have already stated that in the realm of touch, metonymy is the dominant figure: Since touch is a matter of extended bodies, it is to be expected that figures in which relations of contiguity [spatial, but also temporal] are important will play a dominant role. However, looking closer might reveal that what we had naïvely taken to be metonymy is actually a different figure.) The analogical substitution conflates spatial distance and absence in a fundamental figural manner. This conflation is only semantical, at the level of meaning. When one touches a thing, say the paramour's finger, the meaning of the action conflates other parts of the body (for which the finger is an analogical substitute) with the finger itself. But the literal meaning is still that of holding the finger. Metaphors function by denying the literal meaning in a certain sense, which also conforms to the general observation of figural meanings that these are activated, only when the literal meanings does not make sense. Given a sexual intention, holding a finger is not really meaningful, or at least has only a low-intensity meaning, but since this low-intensity meaning does not make sense, it is allocated a high-intensity meaning, and the act is transformed into a sexually charged act. It is not very difficult to find other examples of metaphorical touch.

Metonymic touch is more or less self-evident, and it would not be necessary to spend too much time on it. Our preferred example is that of the dalit's shadow. Why is the dalit's shadow a contaminating touch? It is possible to state that there is no touch in fact, if by touch we mean one person touching another. However, the meaning of touch cannot be restricted only to people touching or not touching another. One can be touched by a shadow, just as one can be touched by the light from a god, or the aura of a saint. This is a clear case of synecdoche (which we take to be one kind of metonymy, following the late medieval and renaissance European classification of metonymy into four kinds, two of spatial contiguity (part for whole and whole for part), and two of temporal and logical contiguity (cause for effect and effect for cause).[19] One's shadow is a very powerful metonymy because it is spatially contiguous to ourselves (it's always with us) and because we cause it (thus it becomes a tool for inference: If the shadow is here, can the dalit be far behind?).[20]

Allegory, which can quite often be mistaken for a figure of analogy, is rather different from metaphor. Tradition quite often treats allegory as a vastly extended metaphor. However, the relationship between the two levels of meaning is not of analogy but of parallelism, and the distinction between analogy and parallelism must be maintained. Two things may be parallel to each other without ever being analogical. As Walter Benjamin reminds us, the distinctions between similarity, analogy, and relation must be maintained, and parallelism can be added to the list.[21]

Possibly the best example of allegorical touch is the touching of the feet of elders (on the Indian subcontinent), or the touching of sacred objects. When one touches the sacred idol, be it a weeping Mary, or Kali, or some such idol, one is touching the realm of the sacred, the spiritual. This is also to be found in masturbation: The physical action is only very tenuously linked to what transpires in the mind, but there is a parallelism here that makes this kind of touch possible in the first place.

We believe we have given sufficient indication of how the distinction between literal and figural touch operates. It is important to remember that this is also the realm of substitutes.

THE THIRD OPPOSITION: GOOD TOUCH AND BAD TOUCH

It would not be necessary to discuss the words *good* and *bad* in themselves, (a) because the meanings of these words are self-evident, and (b) because it is not the ethical notions that we are interested in, but how in society, touch is seen in terms of these two categories. In short, it is the social practice of these concepts that are interesting, at the risk of seeming to posit and accept some kind of difference or dichotomy between practice and some theory that is radically different from practice, if such a theory were indeed possible.

As discussed earlier, goodness and badness are qualities, not substances or objects. Thus, the third distinction of touch is of a different type than the first two. Touching oneself, touching others, and literal and figural touch involve some kind of substance, though the nature of the substance involved may be debatable. There is no substance to correspond to the words *good* and *bad*. It follows inevitably that these qualities are predicates of the earlier distinctions and cannot be found independently of them. This makes the discussion a little complicated, since we cannot discuss them independently, without turning them into some kind of substances, or semblances of substances.

This distinction also is the most variable in time and space. The values of good and bad attached to an event of touch are determined in the action of touching itself, and it would be difficult indeed to find some general norms for defining the good touch and the bad touch. It is here that the generality and stability of literal touch is of some help; it is the relative stability (within a community of believers like ourselves) of literal touch that will allow some semisolid ground on which to discuss the distinction.

Inasmuch as these are qualities and values, they are far more clearly social and communal than the earlier distinctions. It also follows that these are far more variable and changeable. It is in this realm of changeability that the regimes of touch might be transformed or altered, or fubared. It is to be remembered that a transformation of the values of good or bad attached to this or that kind of touch is not necessarily a consequence of some structural change in the regime of touch. A few changes in the notation of values do not allow us to infer that the regime itself has undergone a transformation. This is demonstrated in the devaluation of touch in metropolitan centers. In a crowded local train, it is impossible to avoid being touched, even if you were the purest of brahmans. A good example of meaningless touch is found here; even if one is touched at the most intimate parts, and by the most intimate parts of others, the touch remains mostly meaningless, unless other intentions are sensed. Looking at this crowded local train, one might be driven to conclude that there is no caste hierarchy in metropolitan centers. However, all that has happened, as is well known, is that the various notations have undergone a change.

Despite our earlier protestations, it is important to discuss what constitutes the good touch and the bad touch at a general level, apart from issues of caste, which are our main focus. In the discussion of styles and tones of literal and figural touch, we have already mentioned some of the qualities that might be associated. We discussed altruistic touch and violent touch. Now we need to discuss touch that generates pain and touch that generates pleasure. Pain and pleasure can be generated only by the literal touch, on the assumption, consistent with our position, that these are qualities of physical and material experience. That the assumption is valid is indicated by what might be called the misfiring or even misprision of intention. A touch that is intended to give pain may not necessarily be followed by pain; it can be miscognized, misread. In fact, a deliberate disacknowledgment of the intention might even reverse the expected notation. For example, to someone who hits you with the intention of causing pain, you could always say, "Your touch does not cause pain but pleasure," thus reversing the notation of qualities. The opposite case is quite easily imaginable. This is a common enough rhetorical strategy, known in traditional rhetoric as the paired figure of *concessio* (I *agree* that no man can kill you) and *metabasis* (*but* I am no man).

All the primary forms of pain and pleasure are dependent on literal touch. In the preceding chapter we have already indicated our reservations about the psychological and psychoanalytical discussions of pain and pleasure. Figural touch may generate some amount of pain or pleasure, but the intensities are not comparable to physical pain or pleasure. As to what are pain and pleasure themselves, physically, a discussion is somewhat unnecessary, since as we have maintained, these are self-evident to all of us. No doubt an ethological discussion of flight-fight mechanisms, with pain as the stimulus that activates them, is always possible. However, until we are in a position to give a detailed and convincing description of how to arrive at the social nature of human existence from these primordial mechanisms, such a discussion is unfruitful in this context.

At the risk of sounding peremptory, we have to say that what is good about the good touch and bad about the bad touch is quite self-evident to all of us and needs very little elaboration and discussion.[22]

3

TOUCH IN ITS SOCIAL
AND HISTORICAL ASPECTS I

THE SOCIAL ORGANIZATION OF TOUCH

In the earlier chapters we have attempted to construct the prism of touch through which to look at society. In the process, we have also established some of the aspects of its intersecting planes and their individual and collective textures. It is perhaps time now to take up something larger, such as society. However, in order to look at society, we should know, even more than what society is, where to find it. All of us know where to find society: We find society wherever more than one human being is to be found—since it is only of human society that we are speaking. This obviously is the minimal model of society and ethical interaction, which has two people face to face. However, it is extremely important not to forget that this is only a model of what we might call a minimal society and therefore possibly a misleading model, if not entirely untrue. We have no experience of this minimal society, of which communication theory and linguistics too have become fond, resulting in a very restricted view of their objects of study. The number three, from this point of view, might seem more interesting, but a model of society based only on this number would be only a little less misleading, since the point is that we only have the experience of living in societies that have thousands of people, even if we have never met them personally. We have indicated in earlier chapters that a model of society built on the conceptual pair of "self and the other" is fundamentally inadequate. This model has to pay a price for the ease of conceptualization that is greater than the ease itself. To build a symmetrical, or even an unequal, relationship between the one and the many, by imagining that there are only the one and the other in the world is an intellectual trick that yields some interesting philosophical insights into the one and the other but next to nothing for understanding the world and society.

We also offer a classification of societies, since there are many societies. It should be possible to arrive at a classification of societies before looking at any particular society by asking the question of the source or sources of value in societies. Value is an abstract notion, and it must realize itself in society in some way or other. These realizations we call emanations of value (à la William Blake). The question of the source or sources of value is a question of what societies might believe in, in general. This prevents the question from becoming etiological, even if within a given society etiology might be mistaken for explanation or history. It is clear that the question cannot actually be a single question: There must be a series of consequent questions.

The first question, the most fundamental one, is whether the source of value is thought to be human or inhuman and transcendental. This question is crucial and allows us to distinguish between societies that are nonmetaphysical and societies that are metaphysical.[1] There are, to the best of our knowledge, no societies in which the source of values is thought to be only human; at the same time, it would be difficult indeed to find societies in which no originary force is attributed to human existence. The question is of degree. Inasmuch as most societies are religious, in most societies the originary force of creating value is predominantly attributed to inhuman transcendental existence. The very nature of most religions requires an inhuman source for human existence and human values (the notion of god/goddess, in singular or plural number). It is crucial to remember to distinguish between what a society finds valuable, and what a society thinks of as the source of value. Presumably, if the source *generates* value, it cannot be the *repository or reservoir* of value, resulting in a strange paradox, for often the source of value itself is thought of as valuable. To think that the source of value is, as it were, valuable because it is the source, or in itself, is Romantic, if not conceptually erroneous. Inasmuch as the source must be different from what it sources out, the source itself cannot be valuable. What is valuable is what it produces, what it generates, what it creates, be it labor, or the language of god in Christianity, or some other entity. Romanticism, primitivism, animism, and many other forms of relating to the world in terms of the sources of the world or the elements in it share this error. We have diverted our attention a little to concentrate it on this phenomenon because it is widespread in several societies, including societies that artists, especially science fiction and fantasy novelists have been able to imagine[2].

Our argument is that a community can be identified on this basis, since at this moment we are not distinguishing between community and society. Further complications can be added to the notion by setting these words in the context of other words such as *culture*. It seems that it would be difficult indeed to distinguish between society and community, because although it is always possible to argue that a society comprises several communities, the logic by which these variegated communities form a single society must necessarily remain unclear within such an argument. This ambiguity, or overlap, between the two notions or meanings of words is quite clearly present in the Marathi usage, in which the word *samaaj* may be used both for society and community—in fact, it is used for one's caste as well.[3]

These words bring us back to our chief concern, which is that of the social organization of touch. This is a somewhat misleading phrase, since what we are talking about are the regulations on touch which generate communities which are thus regular and regulated. We have argued that societies could be classified by the principle of attribution of agency of value to human or inhuman and transcendental sources (always in varying degrees). It seems that it is possible to rewrite the debate about determinism and free will in these terms, along with several subordinate debates in theology, Christian and non-Christian.

Those societies in which the source of value is taken to be inhuman[4] and transcendental we call societies of inheritance. If the source of value is non-human (which also means that human beings cannot create fundamental value), then human beings can only inherit it. One only inherits something that exists before oneself. It can be seen quite easily that inheritance accounts for major parts of life in society. Theories of racial memory, or social memory, theories of history—all are grounded in the fact and principle of inheritance. Beginning with one's physiognomic attributes, through language to the social organization in which one is born, the dominant part of one's growing up is spent in learning what one's inheritance is. There are elements of acquisition as one can see: One may choose consciously or unconsciously this or that particular line of inheritance, under the heading of tradition (we remember here T. S. Eliot's assertion in his well-known essay "Tradition and Individual Talent" that a tradition has to be acquired). That inheritance dominates over these minor acquisitions is seen in the fact that although one, like a scholar, may choose to inherit (which is a kind of acquisition at a lower level) a vast amount of knowledge, the inheritance itself is always larger than what one can acquire. In societies of inheritance there are limits on what and how much one can acquire. In a fundamental way, this is also a question of what kind of power and how much power are attributed to human agency. In the religious context, the question of salvation or attainment of heaven is, after a certain limit, beyond human agency, since it is usually God's mercy which will graciously grant or not grant a human being residence in heaven. Naturally this is more true of Christianity than other religions. An interesting point emerges here. If in the more or less Augustinian interpretations of the Bible and of the religious life, it's only God who has the power to save, in Hinduism the question of human capabilities is settled a little differently, although with similar results. There are any number of stories within Hindu mythology that tell of this or that sage who studies all the available knowledges, and does penance for countless years, so much so that the God of Gods gets worried and sends usually attractive women to disturb the concentrated labor of the penitent, which is also a matter of celibacy. Once disturbed, the sage cannot attain his goal, whatever that might be. Or, through a curious mechanism, a godlike figure, or a hero, might accidentally or by design, kill the sage. In any case, the gods seem to have the power to deny, or to grant.

The reader will have noticed that until now we have not brought up the complicated matter of gender. It seems that the matter of gender can appropriately be brought up in a discussion of inheritance. It is quite a revealing facet of human existence that one inherits one's gender, and it is not even a situation of take it or leave it. Questions of caste can be made into questions of leaving it, but the question of leaving, or for that matter, taking one's gender do not really operate in society, except in radical "trans-" groups of what inheritance has taught us to call "men" and "women," or various modifications of this classification and nomenclature. The relationship between gender and caste, as the relationship between gender and class is complicated to the point of incomprehension, as far as we are concerned. It is entirely within the range of possibilities that philosophers or sociologists have some simplifications or complications to offer; however, we are not about to offer any of these. The question we are attempting to articulate relates to what could be called the mutual confusion between chronology and logic. It seems that there can be no question about the chronological contemporaneity of gender. One inherits one's gender at the same time as one inherits one's caste.[5] The issue is, as always, whether the syntactic pre-positioning of gender and caste, so to say, amounts to a semantic priority. The ethical and political implications are tremendous, the moment we make up our minds on this issue: If we decide that chronological priority is equal to or the same as or indicative of logical priority, then our political labor will have to be expended on issues of gender and caste rather than those of class, assuming for reasons that must appear to be mysterious at this point, that all of us wish to do only those things, because they are the most semantically loaded, politically speaking, especially when semantics, ethics, and politics are thought to be so intricately related to one another.

It is amply evident that gender involves different regulative principles of touch, possibly from the neonate stage itself, and that the processes of gendering use these regulations for making human bodies into bodies of gendered human beings. The temptation here is to tacitly posit, in an idealistic manner, an essential human body that only receives these modifications subsequently. Consequently, some struggles against the contemporary regulations strive to make human bodies ideally genderless. There are parallels with reformist attempts to make human bodies casteless.

It is sufficient here to indicate our awareness that gendering intervenes in the processes of the formation of the regulations on human bodies—regulations of touch—which also form castes. These issues will arise again and again, and we will discuss them again and again as they do. Our intention here is to discuss regulations on touch with special reference to caste and its operation in society, and importantly, the possibilities of the transformation of the regulations that govern its operation.

In what follows we intend to discuss the social organization of touch. We will concentrate on the contemporary organization, rather than reconstructions of past organizations. Every organization is underpinned by some principle, usually single, although in many cases a number of principles may govern and regulate the organization, with the proviso that if more than one principle exists, they must further organize themselves in a hierarchy of relatively larger dominions of governance and applicability. We have attempted to establish that the three pairs of concepts (touching oneself, touching others, literal and figural touch, and good and bad touch) are the organizing principles. We have not been able to arrive at a single principle of organization of touchability/untouchability.

The question of whether the source of value is believed to be human or inhuman transcendental can be approached with a pair of simple concepts, that of literal and figural. This pair, as already discussed, is related to beliefs, truth, materiality, and its allegorical counterpart, ideology. A human source of value will have to be literal, especially when compared to what we have called the inhuman transcendental source of values, which is figural, possibly in the sense of allegory. It is important to remember the difference between allegory and metaphor. In metaphor, the literal sense must be transcended because it does not make sense. In allegory, both the literal and figural meanings make sense, in general and in detail. Neither should the association between the literal and the real be forgotten.

The specific features of societies of inheritance and societies of acquisition need to be elaborated, and we shall now proceed to do so in the next section. It is to be borne in mind that we are undertaking this discussion with possibilities of transformation in mind.

SOCIETIES OF INHERITANCE AND SOCIETIES OF ACQUISITION

In chapter 1, we pointed out how epistemological categories get mapped onto ontological categories, especially when it comes to caste: A matter of practice, caste is determined by birth, rather than by the practice, and the practice of caste is later on explained in terms of physical and physiognomical and psychological attributes of persons born in a particular family, tradition, and society. Categories of doing and saying are mapped onto, transposed onto, categories of being. This is illustrated, for example, in Kumaril Bhatta's discussion in his prose commentary on Jaimini's *mimamsa-sutras* and Shabara's commentary, the Shabarabhashya, called the *tupteeka*. The question under discussion is how does one know that a brahman is in fact a brahman? After stating that marks like the sacred thread, or language, or behavior are but corruptible and capable of dissembling, it is stated that the only true category that establishes our knowledge of caste is the knowledge of the caste of the father. In short, one is indubitably a brahman if and only if one is born into a brahman family, of a brahman father.[6]

This tight and uncompromising fit between caste and birth, between caste and being, cannot be undone, or even disturbed. This does not naturally mean that caste hierarchies and injustices were not questioned or cannot be questioned within this framework of inheritance. But it is to be seen that even today, there are existent configurations of this framework that are used within and without the dalit movement. The most important of these is the notion of authenticity, dependent on which is the question, for example, of the right to represent the dalit. As is well known, the question of the right to represent is a matter of great anxiety, rancor, and, possibly in that order, discussion. The tight fit between caste and birth is without doubt the core of caste hierarchies and generates and reinforces at every birth the various typologies. This fit is, in fact, a significant feature of societies of inheritance. In societies of acquisition, birth within a particular family does not in principle confer any special rights, positive or negative. Obviously, the very notion of equality is dependent on the notion that birth in a particular family does not confer, de jure, any special rights (though as is always the case, de facto, special rights positive and negative are actually conferred by the fact of birth).

It seems that this fit is one of the thicker strands in the complex web of caste questions. It is also clear that this is a metaphysical fit. Although it is tempting to say that metaphysics per se is the department of the brahmans, it would also be incorrect, since there must have been a metaphysics of the *shudras*. Inheritance is a matter of history as well, especially in its manifestation in the form of tradition, which is imbibed and internalized and handed down. It is at this point that things become a little more complex, since present-day practices, new practices, not only require legitimation from tradition, but in fact are passed off as tradition. Something that is newly acquired is fervently believed to be traditional. This tropology too needs an examination. In this tropology, the notion of inheritance itself becomes a source of value, in the sense that objects and practices and values that are inherited are believed to valuable because they are inherited.

Societies that are called feudal come the closest to what we are describing, yet we believe that the nomenclature we have devised, "societies of inheritance," is comprehensive. The concept is fuzzier than that of "feudal societies." This fuzziness has some methodological advantages, one of which is that the chronological and, indeed, historical sequencing that is associated with the term *feudal* is not associated with our term. This allows for a mixture of societies of inheritance and acquisition more easily than the Marxist notions of the various modes of production, which admit of mixtures only at the cost of disturbing the conceptual and historical differences among the various modes of production. More or less by the same token, the term *societies of acquisition* comes closest to the term *capitalistic societies*, but it is larger and fuzzier and allows mixtures more easily than the purer term *capitalistic societies*. In addition, the term *societies of inheritance* does not need to posit, while talking about Asia, the Asiatic mode of production, and is thus more sensitive to historical differences between European history and South Asian history. This notion also helps us avoid the temptation to use a scheme meant to explain European history to explain Indian history. We can safely assume that just as present-day capitalism in India is different from present-day capitalism in Europe, feudal societies in India must have been different from feudal societies in Europe. We believe that this is useful while discussing Indian history; otherwise, we would have to talk awkwardly about "feudalism" in India, along the lines on which some historians talk of "medieval" India.

In societies of inheritance, almost everything is inherited, most certainly principles. It is within these societies that the tradition of textual commentary is most clearly seen in the realm of textual studies. In the realm of labor, the potter makes the pot and exchanges it in its commodity-nature and therefore could be said to be the producer of value; nevertheless, the profession itself is without doubt an inherited one. Within the caste hierarchy, there is little possibility of change of profession, and inheritance could still be said to govern the production of value. In any case, it is a moot point whether the potter thinks he and his wife are producing value or merely performing their assigned, inherited tasks. In the realm of literature and other arts, this manifests itself in genres such as tragedy, comedy, epic, and romance, with their own norms. For example, the ancient Sanskritic norms specify that a *mahakavya* (epic) should contain most of the meters and most of the passions and so on, and that there should be no representation of death, defecation, and other bodily activities of a similar kind on stage in drama. Typologies of heroes and heroines come about and typologies of passions, the dramatic and poetic techniques to arouse them and so on. In societies of acquisition, what Bakhtin calls a "concern with the open-endedness of time"[7] is prerequisite for the very notion of acquisition, and the genre that is clearly associated with such societies is the novel.

In terms of religion, in societies of inheritance the question of conversion or protestantism does not really arise.[8] Conversion is a matter also of acquiring new religious principles, and this cannot be done in strict societies of inheritance. It is only in societies of acquisition that conversion is possible; moreover, atheism and antitheism are possible only within such societies. Thus, in societies of inheritance, the notion of the future is tightly bound with the past (the future is a return of the past through the detour of the present, which present, in any case, is concerned with the preservation of the past in its customarily desirable aspects), whereas the notion of time in societies of acquisition necessarily must be open-ended, in the sense that the present is open-ended, to use Bakhtin's phrase again. The notion of acquisition itself must presuppose a future maximally or minimally different from the past, and the past relegated to something undesirable. The dominant myth of progress reinforces this view of the past. It is to be noted that all movements for social change are involved in this view, from the bourgeois revolution to Marxist, feminist, dalit, or some other version of social change.

In societies of inheritance there is less anxiety about values than in societies of acquisition, since the inherited structure and relations of value are not only accepted and consented to but are rather firmly in place. In societies of acquisition, the question of which values to acquire and which not to acquire itself generates anxiety, debate, and possibly violence.

To speak in the manner of Northrop Frye,[9] the dominant mythology in societies of inheritance is that of religion, whereas in societies of acquisition it is that of progress and development.

To sum up the various features of these two kinds of societies, it is actually sufficient to state that in societies of inheritance one's birth determines everything—one's actions, speech, practices of value are firmly mapped onto one's being—whereas this determination, as it were, shifts to that of doing in societies of acquisition. Material production itself begins to be a mark of value, under the aegis of the notion of progress: The more material production there is, the more value is thought to be produced. A parallel shift takes place in the attribution of literalness and figurality to phenomena. That which is thought be literal in societies of inheritance, such as religious texts, comes to be thought of as mythological and figural. That which was thought of as figural (at least figurable), for example, the human body, comes to be thought of as literal. Knowledge, earlier thought of as revelation or understanding of god, shifts to knowledge of the world, shifts toward what modernity recognizes as science. Knowledge, earlier a matter of inheritance, becomes a matter of acquisition and shifts its base from places of religious worship to universities (this is also a shift in regulative principles and practices).

Value as such needs representations, manifestations in lived reality. It needs signs, symbols, expressions. In a society of inheritance, value, in its various forms, emanates from the following: from the king (the explicit political, the various conspiracies, plots, alliances, and so on), from god (in the ontotheological sense, as it were), from the father (in the familial sense), from the male (in the sense of gender), from tradition (in the historical sense, since the sense of history as such is isomorphic with the sense of tradition in societies of inheritance), and so on. It is not necessary to posit a complete inventory here. These, we suggest, are the forms of emanations of value and its perception. There is always an implicit political aspect, but that seems to be a modern idea, a twentieth-century idea, in which time it was possible to say that everything was political. These forms of value have literal contents—for example, the individual who inherits the function of representing the king, the priest's son who becomes priest after his father hands over the job, the younger brother who becomes the patriarch when the older brother dies, and so on.

A few words about birth are in order here. Birth and death possibly represent physical and metaphysical mysteries, without which metaphysics itself ceases to be interesting. However, from our point of view, human life becomes ethically interesting only after subtracting these mysteries from consideration. This is especially important in the context of caste, which is so firmly predicated on birth that the predicate becomes only an analytical extension of the person, rather than a synthetic attribution.[10] Inasmuch as birth and death are primary mysteries for metaphysics, a questioning of caste becomes well nigh impossible once caste is predicated onto birth, because birth itself is a mystery. The mysterious firmness of caste has one of its origins in this predication. It should also be noted that inasmuch as birth and death are natural phenomena, nature as a notion itself is extremely important in societies of inheritance: not just nature in the sense of things that were not created by human beings but also nature in the sense of human nature, which further can be divided according to natural endowments, talents, and genius. Each person can now be classified in terms of his or her nature. A whole quasi-mythological system and hierarchy can be now be erected, to put each human being in his or her place. Furthermore, because it is thought to be natural, it is removed from the realm of human manipulation and transformation. This is an important observation. In contrast, in societies of acquisition, this mythology of classification of human beings shifts to more "cultural" activities: Human beings do not now have naturally endowed abilities and potentials, but only are thought to be artists, mechanics, tradesmen, and craftsmen, and so on "by profession," This places caste on the borders of the realm of human manipulation and transformation.

It is here, in the confusion between birth and natural endowments and authenticity, that dalit metaphysics (including the anti-caste activists' metaphysics) is complicit in the maintenance of caste, inasmuch as this metaphysics too, to some degree or another, needs the notion of authenticity, in which dalit authenticity is available only to those who are born dalit.

The relationship between this conjectured metaphysics and the known brahmanical metaphysics needs some description. It is historically correct to suggest that dalit metaphysics did, in some way and to some degree, question brahmanical metaphysics. Sufficient examples can be collected from what is known as the *bhakti* movement.[11] There also were present in history the various *shramana* movements, especially in the south of India, which were also more or less anti-brahmanical. *Bhakti* movements and the various poets who write in that tradition are by now fairly well documented, from the point of view of caste and gender, and need no further comment here. It is our belief that even though some of these movements and schools of thought must have questioned the otherwise unquestionable fit between birth and caste, the questioning was never really strong enough, and the tropology mentioned above continued to operate throughout this part of the history.

However, before we can attempt to give a historical account of the transformation, we must also consider the principles involved in the transformation. One of the first things to notice is that the very notion of society is divided by caste considerations. Thus we could state, albeit as something of a clarifying exaggeration, that there is no single society but as many societies as there are castes. This is clearly indicated in the Marathi usage of the word that is normally thought to be equivalent to the English word caste: *samaaj*. The word is used to mean society, as in the word for sociology, which is *samaajshastra*. However, there is a more common usage of the word in which it indicates clan and caste groupings, and with increasing modernization in the more urban centers, also class groupings. Thus we have, for example, *teli samaj, koli samaj, kasar samaj*, and so on. Banks, credit societies, cultural organizations, and various other activities and organizations proudly bear names of particular "societies." It should be noted that there are caste-denominated chat groups on the internet as well. This is a matter that sociologists will be more capable of dealing with. We are not in a position to posit anything more than a clarifying exaggeration that what we have on our hands is actually a society that is *circumscribed* by caste considerations, and not necessarily a society that is *divided* by caste considerations, *subsequent* to a prior unity.

There are tremendous stakes involved; the greatest is that of the following question: Is there a notion of society operative that is capable of encompassing all existing human beings? It is clear that the answer must be in negative. If we confuse the notion of society with those of nation, culture, community, *samaj*, clan, and family, could we say that we live in a society? If such confusions are possible, is that not an indication that the notion of society either has become obsolete in practice or is something that is yet to come? This is *not* a question in sociology. No doubt sociology as a discipline has some answers to offer to the question "What is society?"; the question is whether the answers that might be offered are merely conceptual or are operative in everyday life. It is worth remembering the ambiguity in the sentence that Emily Dickinson used in a poem: "The soul selects her own society and shuts the door."

In the context of caste and regulations on touch, how does one understand society? Or, for that matter, purely formally, how does one generate society from touch? There are several solutions to this intractable-looking problem, if we treat this as a purely formal problem. One might consider, for example, occasions of touching, remembering that contact is a fundamental element of touch. its content as well as its form. Another approach would be to wonder why there are regulations on touch when it is precisely caste that precludes the possibility of touch? What are the regulations here, and what is the practice? Are such regulations actually de jure, or de facto? Attempts to answer these and many other questions will lead to an altogether different notion of society itself.

To take a completely materialist position, the necessity of touch would arise only when a particular person, family, clan, or community finds itself in a position of inability to provide for its own needs and desires. It is only then that one would be required to touch others, as person, family, clan, or community. The Marxist notion of exchange then could be used to understand occasions of touching. The subtle assumption in this point of view is that self-sufficiency is a desirable ideal. It is our contention that surreptitiously the notion of society as a collection of individuals—or subjects—operates in our discourse. We still are trying to learn ways of conceiving society in a way that tells us of the formation of the individual, rather than the formation of society from individuals. This is not to suggest that some thinkers have not, perhaps from the nineteenth century, attempted, through incisive analysis, to break out of such an understanding. The question is whether that different understanding is commonly shared by a large number of people: Had it been, these thinkers would not be seen as radical and exceptional.

The discussion of kinds of touch is now of some assistance in identifying the occasions of touch. We are talking about literal touch here, and we have talked of altruistic touch, and sexual touch. These two kinds of literal touch indicate the occasions of touch as well: Sexual need and the need of altruistic conduct are the primary occasions of touch. We have also indicated the contraries and contradictories of these in sexual exploitation and sexual or other kinds of violence.[12] It is from these kinds of literal touch that sociality as such (the practice of living with other people) is generated. Inasmuch as these kinds of touch can be directed at oneself as well, they also yield regulations on touch as they operate on oneself. We have already attempted to establish that altruistic touch is possibly more important in generating and establishing sociality: It is not necessarily only through the dynamic interaction between members of a different sex or for that matter the same sex that something like a family is generated. Within a family it is the occasions of altruistic touch that far surpass the sexual touch in quantitative importance. We should also remember that for our purposes it is not actually necessary to explain how a society comes about, whether through the dialectic of the two sexes, or through the dialectic of individual or, for that matter, collective labor. Society is something that we have inherited for a long, long time indeed, and explanations of the origins of society are only as interesting for us as explanations of the origin of language. Such explanations are complicated by the implicit assumption that first there are individuals who dialectically or through some other historical process come to form society. It would be more appropriate, to attempt to explain the evolution of individuals from society than the other way around. This is relevant also to our discussion of inheritance and acquisition, since the society in which one lives is an inheritance, whereas one's individuality, and even selfhood, is something that is acquired: Selfhood here is the variable and society the constant.

Having made this incidental point, we now continue the discussion of occasions of touch. Here, it is the sexual and altruistic touch within which we locate sociality. Among all the occasions of altruistic touch, that of touching a neonate in order to care for it deserves more attention. An unattended neonate generates strong response, because it is the image of total loneliness and incapacity; at the same time, it is also the image of a larger number of possibilities than available to the bildungsroman of attended neonates. Both these possibilities walk paths that converge on that "possibility of impossibility," death.[13] It seems to us that abandoned infants and children, cared for by strangers, have the greatest interest in acquisition of altruistic touch, and they are therefore possibly the prime location for acquisition as such. It is clear that in the case of abandoned children (and we once again mention John Boswell's excellent historical account of the abandonment and subsequent nurture of babies and children), who have an intense experience of solitude, the experience of their singularity too must be equally intense. Therefore, it is possible to hypothesize that solitude, acquisition, and individuality are related to each other in some way. Strongly acquisitive human beings, in all fields of activity (for example, intellectual acquisitiveness), have a strong experience of solitude, along with a strong sense of their own individuality, and possibly attendant responsibility. The pathos in the figure of the solitary artist, the solitary scientist, and the lonely saint or social worker, working without any altruistic support, gains its power from this confluence. We suggest that solitude, understand in its primary form as missing altruistic contact, is a strong experience that impels beings toward acquisition.

The thematic of solitude is powerfully explored in modernist literature, and it is this which explodes fully into the notion of society as a collection of always already atomic individuals. It is evident, it seems to us, that this operation of solitude and the various deployments of it in literature, philosophy, and other fields of intellectual activity, are a special feature of societies of acquisition.

In a society of inheritance, in its strongest form, there are very few occasions of solitude and, correlatively, stronger forms of community. This allows us to define a community as a particular regime of altruistic touch. It is fairly evident (though we are subscribing to what might even be merely a myth in societies of acquisition), that in societies of inheritance communities are strongest. They are so strong in fact, that within the community, community and world might even come to be synonymous. A question that arises is: Who draws the boundaries of altruistic touch, which are redefined as community, and how are these boundaries drawn? The question of the *who* cannot be posited in isolation. Within a field populated by several statements, within a discourse, loosely speaking, it is necessary but not enough to ask who is authorized to make what kind of statement regarding which thing. It is equally important to posit the question: For whom is the statement made? It is necessary but not enough to ask who benefits from making a statement, but we must ask also, who is supposed to benefit by receiving the statement in its supposed semantic neutrality and plenitude? The methodological benefits in this procedure can be demonstrated quickly by a trivial example. Intellectuals who posit certain ideals are often accused of hypocrisy, inasmuch as they themselves do not practice these values. It is a charge commonly heard from nonfeminists and antifeminists that feminists themselves do not really practice equality among women; dalit thinkers and activists struggling against caste are shown how within dalits themselves there are internal hierarchies, and so on. The claim to power is clear in the statement; it is only certain kinds of people who make such a statement, and the benefit to them is clear enough.[14] However, these statements are addressed only to these particular elements in the field, and not to any other. Those who make these statements quite likely do not address the similar statements to the state, which promises one thing and delivers yet another, or for that matter to the police, who can clearly be seen taking bribes (in some cultures legalized as fines, in some others a plain bribe) from people who break traffic rules, for example. This procedure allows us to locate the user of the utterance not just as the one who profits and benefits, but also as one element within the field, not a terminal point, but one point on a chain, entangled in a network of relationships.

Given this, while answering the question of who draws the boundaries between communities and how they are drawn, we should also supplement these questions by a third one: For whom are the boundaries drawn? We benefit here from the ambiguous nature of boundaries: They keep us in, together, as much as they keep others out. Boundaries have recently been given attention for their exclusive features, and attention has been given also to the breaking of boundaries, or crossing them, and so on. Our argument here is that drawing boundaries is an internal altruistic need and anxiety as far as touch is concerned. Boundaries come about not merely as some reaction to the existence of others (as is presupposed in the view that boundaries are exclusionary only), but as an internal requirement of a community itself, not as an act of self-definition merely, but more fundamentally, as an act of anxiety about lack of altruistic touch. The fairy tales in which children are forbidden to cross the village boundary, or early maps with "here be monsters" written on them, are indications of this internal necessity. This is, as stated already, as much a matter of anxiety, worry, and fear as it is a matter of exclusionary practices. We are speaking of boundaries in both the restricted sense of control over certain areas and knowledge of them, and in the general sense of practices of exclusion.

It might appear as if we have taken away something from the valuable notion of otherness. It can be observed, however, that societies have only a confused and vague notion of otherness. The possibility of the lack of altruistic touch transmogrifies others into something to be feared and hated, into enemies; compared to this transmogrification, exclusionary practices seem a little tame. Inasmuch as the sense of community is bound up with the sense of possession of territory, and possession of knowledge of that territory, sociality as such is bound up with these. Our argument is that primary sociality is dependent on altruistic touch and its regulation. The possibility of the lack of altruistic touch is transmogrified into the possibility of hostile, violent touch. As is well known to students of nations and nationalism and communities, this transmogrification is a strong alchemical element in the unity of societies and communities, which admits even of a dialectical interpretation. In terms of parallels, the derivation of hostile touch from within altruistic touch might seem parallel to something we have already castigated by implication: the derivation from the self of a singular other, qualified by the definite article, and yet capitalized in order to indicate its generality ("the Other"). However, that parallel is invalid here since we are talking about societies and not an individuated subject. We have already argued that it is entirely misleading to think of the unity of societies, literally or metaphorically, in terms of the self, although nowadays nothing is more popular than this deception. What we are attempting to show is the anxiety which makes us all huddle together when the clouds burst, the lightning cracks across our eardrums, and our children cling to us, or when a tiger is roaming about on the boundaries of the village. The experience of hostility is, we argue, in its primary form, the experience of anxiety produced by the possibility of the lack of altruistic touch. For the same reason, we should not confuse this with the possibility of the lack of mother or father. With these remarks we must return to our more immediate concerns.

We have remarked above that solitude is a somewhat uncommon experience in societies of inheritance. These societies now belong mostly to the past (although we attempt later to demonstrate that under the heading of "modern" society there is a mixture of these two types of society). Therefore, we take our examples from more or less hypothetical knowledge, fundamentally dependent on social and material forms of memory and traces. In fifteenth-century Maharashtra, what kind of human beings would have the experience of solitude? Certainly, not members of a family living in a house, since one would normally have someone else in the house, unless the whole family itself is excommunicated. Humans who travel for some purpose are far more likely to have some experience of solitude: In the bullock-cart on the road, for example, one may have such an experience. But on average, tradition itself tells us, there would be a whole caravan of such carts, and only extremely rarely a single cart. Naturally there would be one or two sword-bearing types who would give an added sense of the promise of altruistic touch (please note here the near total transformation of altruistic touch into defense and its correlative aggression). In such societies, quite often, bathing too is not a solitary activity: One bathes in the river, or in the lake, where some other people would certainly be around. Possibly other bodily activities were performed in solitude, though we are not sure of that either. Neither do towns and cities of that era provide the kind of solitude in public that is available in modern cities and small towns.

There would, however, be a few professions that would involve some degree of solitude. These include spying and message-running. The runner, who relays messages and documents to the next one, would run alone and would also run long-distance. Spying by its secretive nature would also give a measure of solitude, until, as it were, the secret knowledge is delivered to the relevant official or agency. The traveling artistes would move in troupes, and the jeweler too would move with some protection.

How different is the picture of solitude in a society of acquisition. In societies in which acquisition is firmly established as a good value, one begins with solitude, and then acquires friends, colleagues, and other professional or nonprofessional companions. We need not give even a minimal description of this experience, since all of us have it to some degree or other. To a certain extent these observations allow us to say that communal and social existence is primary and the experience of solitude secondary, as far as the chronological changes from a society of inheritance to a society of acquisition are concerned.

THE SHIFT IN REGULATIONS

To evaluate transformations that have taken place and that might take place, we first must identify whether the existing structure of values associated with the poles and their interrelations is capable of transformation. Such an evaluation involves an examination of what a society professes and what it practices, assuming that there is a difference between explicit and implicit values. The difference is clearly seen when there is a transgression of the explicit and professed law, but there is no social recrimination or ostracism. In the various infringements of caste taboos, especially in the sexual exploitation of dalit women by upper-caste men, the regulations on touch are suspended. It almost appears as if this infringement belongs to the structure itself, since each structure must, in some way or other, or to some degree or other, suspend itself from time to time. However, there are other infringements of the same type that would be seriously punished and not necessarily by law. In the earlier period of history before the eighteenth century, there were explicit regulations on touching the lower castes, and presumably they included regulations on touching lower-caste women.

A person who could afford not to respect explicit regulations must have a large share of symbolic capital. This is enigmatically illustrated in a story told by Godse Bhatji in his *Maza Pravas*. We summarize below what he heard in a town on his journey toward home from the North.

There was a brahman, in the town where Godse Bhatji halted, who was well versed, among other things, in the science of eroticism (*kamashastra*). He happened to see the rather young daughter of one of his servants (a woman herself, who probably washed the vessels, and so on, and presumably of a caste lower than the brahman), and noticed that the girl belonged to the type of woman known in the *kamashastras* as *padmini*.[15] He gave money to the servant; presumably he paid for everything that was needed for the girl as she grew up, and brought up the young girl for his own sexual pleasures, and when she was old enough, "kept" her. Their sexual relationship continued for some years. The girl was exceedingly beautiful and soon received the attention of other men in the town. Meanwhile, the brahman's wife began to protest, to which protests he replied by saying that he had given her sons and daughters, and had done his husbandly duty, and was free to express his sexual energies in whichever way he liked. It soon transpired that several people in town were making advances upon the girl, making it difficult for her to walk on the street. She asked the brahman for advice, and he told her that he would not have any objection to other liaisons, but she should take one precaution: She should collect the sacred-threads of all the people who visit her.[16] The woman did that. Eventually the wife complained to the king; the scandal broke; and the couple were summoned to the court. The brahman admitted to what he had done but asked permission to produce some interesting evidence that had a bearing on the case. He then told the woman to bring the pot in which she had stored the sacred threads of all the people who had visited her. He presented the pot to the court and asked people to identify and take away their own sacred threads. When he was asked what he wanted, he requested that he and the girl be permitted to leave town. They were permitted. Presumably, they lived happily ever after.

This instance might serve to show the flexibility of the social practice of structured values. In practice, values that are explicitly stated can be transgressed, which is more to the point. Transgression itself leads either to tragedy, comedy, or romance. Our instance is one of romance. Tragedy, in which the tragic hero's values are only implicitly supported by author, other characters, and audience, is another example of transgression, with comedy serving as an instance in which transgression of values is severely punished in the end, and social order firmly reestablished (in tragedy we have the hero as the sacrificial figure, as the *pharmakos*; in comedy we have the villain as the sacrificial figure).

My mention of literary genres might be thought inopportune; however, we may as well take advantage of some of the discussion of genres that do, after all, relate to values. An interpretation of the action of the typical genres, based on the principle of transgression, allows us to see values that are fixed in society. The regularity of action, and the operation of values in these genres corresponds to the action and regularity of values in society.

We have suggested that it is possible to derive a community from occasions of touch. Let us look at the other picture, that of societies of acquisition. Allow me to take an almost clichéd snapshot of a "modern" city. We revive here the kind of touch that we have described but not used till now, the third kind, neutral touch. It seems that it is necessary to posit this third kind precisely because of the shift from societies of inheritance to societies of acquisition. Think of the crowded city bus, or tram, or, in Mumbai, the local train. There are a great many human beings standing or sitting beside each other, touching each other with various body parts (almost never with fingertips). This is a remarkable picture, since we believe that very few of these human beings experience touch even though they are all physically compressed by each other. It is possible to suggest that the sense of touch is neutralized for the duration of the journey. The experience must be different in different urban and metropolitan centers, depending on what is thought of as touch and what is not, and what are the specific regulations governing touch. In Mumbai, even if whole lengths of bodies are touching each other, from all sides, there is no experience of touch (unless of course there is some kind of sexuality involved).

We can now locate a major change in the organization of touch and its principles of regulation: In societies of acquisition, touch can be shut off, denied. It seems that this is a newly learned ability. Involuntary touch is now possible, whereas in the organization of touch in societies of inheritance, there would be little possibility of involuntary touch, and even if there were occasions for involuntary touch, the touch would be followed by retraction or protraction. It is important to note that the kind of neutral touch we are describing is neither literal nor metaphorical. This kind of touch, especially traveling in a crowded bus or train, is something that is felt strongly only after a delay, when the sense of touch might become active enough for one to feel like "washing off" all those people's touch by having a "proper bath." We have already indicated why the bath is the negation, a washing off of touch. This, as already pointed out, allows us also to see why, in the brahmanical tradition, washing oneself is a "purifying" action. The neutralization of touch, this deadening of the sense of touch, is a modern phenomenon and marks one point of difference from the earlier regime of touch; it could even be the point on which the regime turns and shifts.

It is abundantly clear even without our framework and observations, that neutralization of touch is necessitated by, and probably is proportionate to, the number of people gathered together in the same space and time. The important feature of this gathering of people is that they are not of the same community; neither have they gathered for some festive occasion in which several communities might possibly assert their harmony in one religion. This gathering of people is more a coincidence, as it were, than a gathering of people with some possibly unifying intention. The only value that unites this gathering, this vast and haunting metropolis, is that overpowering modern intention of acquiring more money. It is tempting to believe that caste distinctions as regulations on touch can no longer operate in this close-pressed mass, where someone's buttocks are pressing against your own and someone's elbow is digging into your chest, and if you turned your face even a little this way, you could kiss someone's ear or cheek, and if a little that way, someone's hair, or back of the neck. But a moment's inspection shows us that at this point, caste is as strongly operative as ever: What has happened is that all these people have shut down, suspended, killed their sense of touch precisely in order to avoid unwanted contact. The traditional regulations are operative strongly enough to force people to change their bodily comportment and put off their senses. We have reason to suppose that this de-communalized gathering is a new form of society, in this unique space-time. Naturally, what is temporarily suspended must resume, minimally according to the law of displacement, maximally as violent symptoms of a repression. In short, what looks like a mass of people who are born equal and who suppose themselves to be free, and of a fraternity of economic intentions, is in actuality quite distant from the late-invented Enlightenment ideals. Our comments are directed at those liberal and liberalist thinkers who describe this gathering of people as a "modern" gathering. It should be noted that what we have said applies most strongly to the space-time of the bus journey or the train journey to and from work, early in the morning and late in the evening, from and to Church Gate, or CST terminus. We might also suggest that the intensity of strife among castes and communities gains its edge in urban centers from this forced gathering under the economic banner. It is a fact well known that the nature of communal violence is much softer in nonurban, "rural" villages; in fact, in the worst of the scenarios so far, 1992, there was next to no violence in villages.[17]

Societies of acquisition need individuated and uniform aspirants to consumer power and status, and in the transition from inheritance, they need to break communities down to individuals. This process is necessarily violent as far as societies of inheritance are concerned. Which ideals are invoked to justify the ways of such societies to societies of acquisition, and the more ethically minded human beings, is a matter of historical development: It can be the white "man's burden," or it can be the justifications of empire or the spreading of the word of a personage no less than the singular god of monotheism, or it can be the myth of progress and development. It can even accommodate state-sponsored literacy campaigns and the increasing insistence on computer literacy. Nationalism is part and parcel of this process.

The neutralization of touch is seized upon by the aggressively sexual marketing of commodities, which is assisted by physical health as a moral ideal. This process, which is more or less complete in India, and large parts of the world, usually starts with marketing of soaps, cosmetics, and underwear. We would strongly suggest that the comprador collaboration between soap and bathing is strictly nonfortuitous, in societies in which bathing is an action performed in constricted spaces, within four walls, with the door closed (unless of course, there is a sexual invitation involved). In any case, in "modern" urban centers, this is one of the few space-times when, and in which, a much-wanted solitude is available. There are very few spaces that now are as charged as the bathing space, which Hitchcock was quick enough to see and clearly told us in his *Psycho*, as far as American society is concerned. (Hindi cinema, and we have mainly films produced in Mumbai in mind, deal with bathing in a different way: It is merely a narrative device to uncover women's—and in a few cases, men's—bodies. It is to be noted that this happens in public spaces, or natural spaces like that of the waterfall.) In most societies, bathing is a purificatory action, for reasons already delineated.

The neutralization of the sense of touch is rather important, since it enables us to pinpoint a specific moment in the transition from societies of inheritance to societies of acquisition. Before the transition is complete, there is a historical moment, of a rather long duration, when societies are in ambivalent, if not an ambiguous, state. We have indicated this ambivalence with the word *mixture*. When the transition is complete, the occasions of touch will themselves be clarified into the four types (altruistic, sexual, exploitative, and violent). This is not to suggest that we will be able to locate the various occasions of touch into these clear and almost pure four kinds. Most societies are mixed in the above sense. The very notion of history and historicity will have to be denied in order to suggest that these four kinds actually exist somewhere in that clear-cut form.

This gathering of people also is under the aegis of a new mode of working, a new work ethic, so to say. Let us clarify at the outset that we are not in a position to analyze, consider, or intervene in the various debates about the economic interpretation of labor, the difference between wealth and status, or economic and symbolic capital, or the various possible and actual interpretations of labor and production, and the dialectical interaction between human beings and nature. Our interest is in the social conditions in which work is undertaken. This issue is crucial, because it has traditionally been used to distinguish between caste and class. The business of work is also the business of professions, and we have already written a little about the matter of inheriting or acquiring professions. We have also slyly suggested, in a rather orthodox manner, that chronological priority amounts to semantic priority (one inherits one's caste and gender before one can acquire one's profession).

This new mode of working is firmly and clearly nodular: The very formation of towns and cities is nodular (cities are formed where several interests come together to form a nodule, so to say: There must be a river; there must at least be two or three trade routes that cross each other at this point; and there must be a confluence of religion, trade, territory, and resources as well—it is not merely fortuitous that in Hinduism the confluence of rivers is a very powerful metaphor).

Before we consider this nodularization, let us pay some attention to processes that make up the earlier organization, again through a story that Godse Bhatji recounts. This time it is a story of something that he did himself.

On his way up North, he went to Ayodhya. He did something there that must, for us, be incomprehensible. This is what he did. In Ayodhya he heard that there is a place where if you dig, you find the burnt rice left over from Dasharatha's *putra-kamesti yajna*.[18] This particular one was done in order to have a son, one of most illustrious personalities in Hindu mythology, Rama. Godse Bhatji and his companions went to the place to find people digging up the ground and collecting more than handfuls of burnt rice, presumably left over from the *yajna* that Dasharatha performed, and which gave him the son. As is well known, myth has empirical consequences: Godse Bhatji *does* after all find the burnt rice and brings it home with him, presumably, as do a number of other people. On the assumption that that there could not have been any burnt rice left over from that mythical time in which the *yajna* itself was performed, and that nobody would be so naïve as to fully believe that it did, several interesting aspects emerge. What emerges first of all, is perhaps the power of rumor, which is a form of collective belief transformed into knowledge and truth by that fact of being collectively held. Next emerges the fact of this notion of the leftover (about which the French scholar Charles Malamoud has written in his *Cooking the World*).[19] The social and collective access to powerful, sacred time that this physical and material residue gives and what follows from it is the fact that the time of the *yajna* can be made present again. This representability of sacred time is a very strong feature of all societies of inheritance, in which rituals normally perform precisely the function of re-presenting the otherwise unrepresentable. We have indicated this when we spoke of how ritual literalizes metaphors and allegories.

From within the framework of the myth of modernity and science, the literal meaning naturally is rather different. We would rather think that some clever person has actually dug up the ground, scattered a lot of burnt rice there, and then covered it up with the upturned soil, and then spread the rumor. Please remember that this is 1857, less than two hundred years ago. The town itself probably keeps the secret because it would fetch people to the religious center and increase the revenue of the whole town. Possibly the town itself believes it to be true. But someone knows the truth and keeps quiet about it. On the assumption that it is the increase in monetary circulation which is the impulse, we might be able to say that here we do not have a clear-cut separation of economic emanation of value and the symbolic emanation of value. This might be a feature of societies of inheritance. The physical pseudo-residue of the time of the birth of Rama enables Godse Bhatji (and others) to have a direct access, through a perfectly literal cause-and-effect metonymy, to that time. The immediacy and self-evidence of the access obviates any questions that might arise about the validity of the claim that it is that kind of a residue, rice burnt in the *yajna*. As usual, truth is self-evident, here too, though the framework is quite different from the modern framework. It seems that the process of nodularization is an ancient one and represents cultural processes at a material level. An interesting interaction between silk and religion, for example, is described and examined by Xinru Liu in his book *Silk and Religion*.[20] Naturally, this confluence does not always need a confluence of rivers.

We have attempted above to delineate one process of thought and cultural activity in order to juxtapose it with what could be called a more modern nodularization, which we can witness in metropolitan cities. Needless to say, the modern form of nodularization is premised upon quite a different notion of society, life, human action. It is in societies of acquisition that this nodularized space becomes the place par excellence of solitary individuals. (To put it in other words, the size of personal territory decreases as the number of people increases. It seems to be an inverse proportion, and at its extreme, only the mental realm remains as personal territory, to be staunchly defended.)

One of the earliest symptoms of the shift toward a society of acquisition is the critique of the ritualism of societies of inheritance. What began as a call for reform, not so much of religion as its practice, this critique seeks to weed out religious practices that reinforce the inherited individual, familial, communal, and social values in the cycle of rituals, which could be daily, seasonal, or annual, according to the variety of calendars. Some of the calendars of ritual might even be different from each other, although in most societies, a certain arithmetical and astronomical and astrological arrangement of this cyclicity and its calculability is an important feature of the inheritance. Jotirao Phule in his *Shetkaryacha Asud*,[21] for example, writes a long list of rituals prescribed by brahmans (and in his argument, cunningly forced on nonbrahmans); he argues further against these rituals. It is well known that he did not mind being called a "protestant Hindu," and one obituary in fact calls him Luther. (The organization of ritual, and the necessity of ritual is most strongly felt at times of birth and death, the metaphysical import of which we have already mentioned.)

The mention of protestantism will, no doubt, remind the readers of Luther, Calvin, and other theologians of the period of the Christian Reformation. But protest against ritual within the Christian framework, and the direction of historical change known as Protestantism, must necessarily be rather different from protest against Hindu ritualism and the direction it takes toward social change. We will speak of protestantism within the context of dalit protests, and the movement as such, and it is important to emphasize that we do not mean to suggest that there is a necessary relationship between the two. There is some historical evidence to suggest that there might well have been a relationship: Tom Paine was Jotirao Phule's favorite writer, and the general economic tone of Phule's argument and the notion that there is a priestly social group which exploits others for its own benefit by mystifying the uneducated seem enough to posit a relationship. However, methodologically speaking it would be more interesting to see whether there is anything common to Christian Protestantism and Hindu Protestantism inasmuch as both are protestant, in the sense that they protest against an inherited and rather firmly established social organization, and whether the conditions that make protests possible are similar.

Given the practice of Hinduism in the latter parts of the eighteenth century and the first half of the nineteenth, what form would a protest take? What was the practice of Hinduism in the period? It is necessary to note at this point that the temptation is to reach into the archive and dig out the books that are most frequently referred to—books on law, governance, morality, and literature—and from the sentences in these books to construct our evidence, on the basis of which we will construe the past; this temptation needs to be avoided.

True, the British judges too settled some cases possibly according to the *Mitakhshara* or the *Nirnayasindhu*,[22] and so on. However, it seems to us that in order to reconstruct a religious practice on the basis of religious texts, we must have a poetic ability, in the good and the bad sense. Be that as it may, our concern is with the notion of protest, and the conditions of its possibility.

We will speak of nineteenth-century Maharashtra. One indubitable condition that had to be in place before any kind of protest could take place was a different notion of employment and of education. It is important to appreciate the role that these have played in nineteenth- and twentieth-century Maharashtra. These are rather different from what were thought to be employment and education in traditional Hinduism. The question turns on two points and naturally changes its shape somewhat randomly as the nineteenth century proceeds to its arithmetical close, in 1900. The most fundamental point is of access; the second point is of the kind of access.

The second occasion for a fundamental change was created by the new education. Education in the colonial period has been studied, and some account of the historical situation is available, which we will not reiterate here. What needs identification is the relationship between education and employment, and the difference between an education inherited and an education acquired. The question of the substance of education is also significant, in the sense that whether it is the hands, fingers, ears, or the whole body or the mind which is being educated, together or independently of each other.

THE AXES OF TRANSFORMATION: EMPLOYMENT AND EDUCATION

When discussing employment, it is important to remember the discussions regarding workplace. If the workplace is contiguous to the space called "home," we have a rather different formation of employment; if it is not, and if the employees are to come together in a space that belongs to none of them, we have another formation on our hands. The difference is most clear in Maharashtra, and presumably in other areas as well, in the traditional, almost clichéd image of the potter working at home and the laborer working in a factory.

The question of employment is for a lot of us the most important question while discussing the transformation that caste has undergone. The possibilities of work have shifted from the inherited profession to employment sought out and acquired. In the earlier form, in inherited employment, the relationship between work put in and returns received in exchange was quite often mediated by a variety of symbolic, or cultural, factors and considerations. The exchange was not the "pure" exchange of labor for wages that we use as a conceptual tool to understand economic relationships among people. The method of work too is significantly different, which we must remember throughout our consideration of the past forms of work. The cobbler in Maharashtra, for example, quite likely does not have the abstract measuring system that we now use in the form of inches or centimeters, represented on some length of material. The thread, the palm, eyesight—these are the measures used. Designing footwear necessarily involves having the person present in order to take a foot imprint. In short, the difference in style of work and between working at home and working in a centralized workplace needs to be reiterated before we can really begin to think about the fundamental changes brought about by a different form of earning money.

A few things have to be detailed, as necessary reminders to ourselves, who are more than willing to grant the differences between these two. In the earlier form of work, quite a lot of the knowledge is a matter of skill with hands, eyes, ear. The maker of musical instruments is a case in point. The diameter of the gut-string to be tied to the string instrument is measured by feeling it; there is no instrument, as far as we know, which allowed the *sitar* or *rabaab* maker or the *tabla* or *tanpura* maker to measure the diameter in relation to an abstract ideal measured by abstract numbers. Making an instrument is almost like tuning it: You cannot do it with the help of numbers. There is no mechanical precision tooling. There is no mass production. Each thing, each commodity is designed for each consumer. In his essay on the storyteller, Walter Benjamin speaks of the imprint of the potter's palm on the pot. We take this to be a metonymy for all forms of work, except, possibly, for one of the oldest things produced, fabric. We do not only have in mind the matter of mechanization of the making of things, and the various interpretations and implications of it, but also the matter of abstractions and the experience of making things with one's own hands. Clearly, there is a specific link between mechanization and abstraction, since the former cannot do without the latter.

What is the relationship between education and employment in societies of inheritance? In the case of the brahman, the link is clear: The more educated you are, the better your chances of being appointed judge, priest, royal adviser. In the case of the dalits, education cannot be a question. In fact, education might even be spurned, being a brahmanical activity. In any case, there is little access to education, in the primary sense of literacy, and in the ideological sense of that which the dominant recognizes as education. It should also be noted that education does not make possible a change in the profession. It merely enhances one's value within the inherited professions, even for the brahman.

It is a well-known fact that it was in the nineteenth century, with the coming of the employment opportunities created by the by then firmly established East India Company, that a nontraditional and noninherited employment became seriously available. It seems reasonable to state that the earliest form of education, understood as "training," was in the Native Army. It is equally reasonable to suppose that this was imparted because the way the British thought of warfare was different from the way in which the recruited "soldiers" were used. Now one had to learn to bear the body differently, put on a gorgeous uniform—the more gorgeous the higher you rose in the army. All the bodies are to be borne the same way, and the uniform takes away the visible marks of caste—the brahman's *shikha* cannot be seen under the cap, just as the lack of it becomes invisible. This does not mean that the distinctions are completely obliterated: The cooks are different; the food is different; and the living quarters too are different. Nevertheless, already there is the hint of the possibility of a completely neutralized space-time. It is here that the brahman can become a soldier, along with the castes traditionally associated with warfare. The historian Seema Alavi's book *The Sepoys and the Company*[23] contains some valuable information about how castes would claim to be the "warrior castes." There is historical evidence to believe that the disquiet around caste within Hinduism was first generated—and we are only talking of the colonial period—in the army. Alavi has paid some attention to this in her book (though she is less interested in caste than in other aspects of the relationship between the *sepoys* and the Company). The constitution of the Native Army, initially by accepting and then possibly encouraging the existent mythological classification of people into varying degrees of martial potential—something we have discussed while talking of inheritance—is significant because this possibly represents one of the first locations where people of differing castes had to come together in the same time-space, possibly stand beside each other in the parade, and possibly even fight whoever was the stipulated enemy, like brothers-in-arms. This is more true of the northern and eastern parts of India, though, we must remember, that the Company's army which fought against the Peshwas in the various battles had Maharashtrian and other soldiers. Those who served in the army were economically far better off than those who continued to insist on the rhetoric of land, soil and inheritance and attempted to cultivate it. Soldiers could rise higher, earn more money, and return, on retirement, to purchase more land than their father had. Especially for the dalits, the service in the army must have been an instructive and an empowering experience. When the Mahar Regiment[24] was disbanded after the First World War, a few years later, the *mahars* were to make a petition to renew recruitment in the army. This part of the history, and role played by people earlier employed in the army in the anti-caste struggle in the twentieth century, especially at the Mahad *satyagraha* is well known, and we need not reiterate it. The disturbance in caste hierarchies and the inherited social organization in these relatively new spheres of employment needs a very detailed historical identification and analysis.

In the first few decades of the nineteenth century, various Christian missionaries had set up schools, and the general disquiet that several intellectuals felt was translated into the demand and establishment of other schools as well. The history of education in this period is well known, and we will not even summarize it here. In any case, what is of interest to us are the changes wrought by these new forms of employment and education.

An interesting confusion is instituted by these new forms. The more or less emancipatory values of education are confused with the values associated with economic upward mobility. This needs some consideration, since the kind of education we are discussing is also connected to employment. The instrumentality attributed to education now only ensures that a certain amount of embourgeoisement of the uneducated will take place. This has implications that become clear in the struggle for employment in the later decades of the twentieth century and into the twenty-first, in which the promise of economic uplift of the dalit is offered as the solution to the "problem of caste." The confusion consists in finding economic solutions, such as they are and might be, for what are ethical problems. The emancipatory attributes of education can be diverted or perverted so that they do not disturb the social organization. It seems to us that this is a mark of societies of acquisition, in which acquisition of economic power becomes in itself a value. This also allows anti-caste struggles to be diverted, in a certain sense, to issues of reservation, the bargaining over its amount in percentage,[25] its enforcement, and so on, in which the judiciary finally becomes the sole arbiter. In the meantime, education itself can be neglected a little, both by pro- and anti-caste policymakers, leaders, and activists. The process of this diversion is slow and becomes apparent only when one compares the educational activities of Phule and contemporary dalit leaders.

It is worth noting in passing that in this period (the nineteenth century especially), education has differing implications for women (usually upper-caste women) and for the dalits. The confusion that we pointed out is not really available to women, since for the majority of women, education retains its emancipatory value, rather than getting reduced to a mechanism that enhances job opportunities. This is particularly true of women of the "upper" castes. Women are educated by fathers and husbands not because it enhances opportunities of employment, but because it is spiritually edifying and, now, morally right. It is a matter of the spirit, rather than employment. The churning up that took place "inside" the home because of education, how it became possible further to domesticate women as the protectors of purity and other such higher values, especially of spiritual values, has been studied and much discussed. We believe that the churning up of the "outside" by the increased opportunities of education and the entry of educated dalits into the job market has to be studied in some detail still. It is important, for example, from a feminist point of view to study the dalit home and the effects of education on it. If dalit poetry and autobiography are anything to go by, a certain pattern is discernible by the middle decades of the twentieth century. Usually the parents undergo extreme suffering to educate the child, and usually it is the mother who makes the most sacrifices, possibly beginning and not ending with the sacrifice of food itself. This is particularly true, frequently, of the family of the landless laborer or the urban nonorganized laborer. The educated child seeks employment in state-sponsored enterprises, especially the administrative bureaucracy, or in the field of education.

The confusion between education as enlightenment and education as mere prerequisite for employment is a major achievement for societies of acquisition. Society can now pretend that by finding or positing economic solutions to problems that are both economic and ethical, it has solved the ethical problem as well, and the need for enlightenment, which the upper castes need as much as any other, can be ignored. Education, in this sense of the word, is reduced to a rung in the economic ladder (which by now everyone is supposed to climb upward), and not a question of knowledge or even a skill learned for its own sake or a rung in the ladder of enlightenment. That this is the case can be seen in how the emancipatory aspects of education have been neglected in society. This is true for both the social groups: The one that has traditionally been used to some form of education and the group for which education still is something new.

Education institutes a difference between being and doing as far as the educated dalit is concerned, but the strong determination from birth still operates, which helps us understand how the above confusion contributes to the maintenance of the traditional determination of caste by birth. The traditional mythological typology of classification of people still operates, even if the dalit now is doing things that are rather different from the inherited doings. The relative independence of the spheres of the economic encoding of value and the more "cultural" encoding values understood especially in the typology of people is discernible here and might help us understand why these encodings are retained by societies of acquisition: As long as the economic acquisition goes on undisturbed, the other spheres of activity can be safely ignored or integrated within economic values and actions.

The use of education as a mere rung in the ladder of economic upward mobility underplays the transformation of modes of thinking, especially the analytic ability. In societies of acquisition, economic activity has its own mythology and value. The myth of money is the dominant myth, and several explanations of life and phenomena end by arriving at this apparently deepest level. One might discern here some implications of what Walter Benjamin attempted to articulate in the small fragment called "Capitalism as Religion."[26] Earlier we attempted to argue that to construe the economic behavior of people as the last explainable level of materiality, as the final instance, is somewhat inadequate, since there are things below that level as well. Our notion of materiality, therefore, is different from what we have called the metaphorical aspects of the notion of materiality. In societies of inheritance, especially in the completed forms of religion that they have, the construal of economic activity is relegated to the mundane, and even the vulgar. In societies of acquisition this evaluation is mostly inverted. This evaluation is related also to issues of literal and figural meanings, something that we have already discussed in some detail.

With this brief sketch of the differences between societies of inheritance and societies of acquisition, and the lines along which a shift in regulations on touch takes place, we are in a better position to understand the operation of caste in contemporary societies of acquisition. What is of crucial importance is the realization that these two kinds, or genres, of society that we have posited are most frequently to be found in a mixed form, and not in their hypothetically "pure" forms. In our view, the shift begins in the first half of the nineteenth century.

4

TOUCH IN ITS SOCIAL
AND HISTORICAL ASPECTS II

THE SOCIAL REGULATION OF TOUCH II

In the preceding chapter we delineated some of the social aspects of the implications of the social organization of touch. We need to understand now what this social organization is, as such. There are two ways of doing this. One is to study the documents that record and formally prescribe these regulations, actively and passively. We refer to these documents as injunctive documents. The other is to look at society itself and see what regulations are operative. Our view is that the first is not really useful, for a variety of reasons, the first of which is that these regulative and prescriptive documents are brahmanical only and can give us a picture of the complete society only through an inverting interpretation. The other reason has to do with the relationship between text and historical events, and the narrative mediation between these, even if we are working with "uninterpreted" primary documents. Along with this specific problem in historical methodology, there is also the other equally methodological problem of working with dominant texts in order to construe the life and regulations of the dominated.[1]

Although by no means do we want to suggest that observing society is easier than historical reconstruction, it does seem to be fraught with only the problem of point of view. Facts in the present are easier to handle than facts in the past, even though both are construals, and not unadulterated facts, since there is less of a burden of memory on present facts than on past ones.[2] Moreover, the operation of these regulations on touch is a matter of everyday practice, which is not really amenable to historical reconstruction, although several historians have attempted such reconstruction.[3] We attempt a middle path: to work with generalities and schemata, with attention to the practical operation of these regulations in contemporary society.

We have semiformalized into genres societies characterized by inheritance and those characterized by acquisition. In the contemporary world, however, these genres manifest themselves as stylistic tendencies and elective affinities, filiations, and affiliations, to use a phrase from the writings of Edward Said. That there are strong stylistic tendencies toward societies of inheritance can be clearly seen in the difference between the generally European-American form of capitalism and capitalism as it is found in India. Before the rise of capitalism (mostly in the nineteenth century), the dominant form of society was presumably that of inheritance. It is important here not to think the tendency as a persistence of inheritance or a residue of an earlier form, because that assumes that the present form is progressive, and that there will be a time in the future when these undesirable residues will die out like dinosaurs. This inability to think in terms of progress possibly is the advantage in working with the notions of the two genres of society, that it disallows us to tell a story of a historical progression of society from one to the other. Attention must be drawn also to our use of the term "the nineteenth century." This is the period that is most commonly referred to as "the colonial period," and we too have used the term occasionally. However, it would be somewhat misleading to cover up the whole century by the heading of "colonial period." Moreover, the use of the term "colonial" looks at the past of that period with the lenses of "precolonial" and, even more significantly, looks to the future with the somewhat newly acquired lens of "postcolonial." This sequencing is to be avoided, we believe, because in it the term "colonial" sits in the middle in an unqualified and innocent manner, a base noun from which other nouns are formed by the supply of different prefixes. Therefore, we have chosen to count with numbers rather than adjectives that are made to behave like nouns.

It would be difficult indeed to observe the whole but open-ended society,[4] and we need to divide the problem in its generality. We have already suggested that contemporary society has the two stylistic tendencies of inheritance and acquisition. Inasmuch as the tendency toward acquisition is strongest in metropolitan centers, we suggest that occasions of touch in metropolitan centers are different from occasions of touch in nonmetropolitan areas, which tend toward inheritance.[5] This distinction comprehends the distinction between urban and rural social organization. We do not make a serious distinction between small cities and metropolitan centers, or small cities, villages, and hamlets, since we are attending to the stylistic tendencies rather than the social specifics.

When we talk of the social organization of touch, we are talking about the organization of the meanings of the kinds of touch that we have already discussed: That is what is being regulated and organized. Now we must also consider the substance, that is to say the bodies on which these regulations operate, and the difference that different bodies make.

It is a remarkable phenomenon that the maximum regulations actually operate on the body of the brahman. These regulations are mostly purifying in nature, and that allows us to mark the vulnerability of the brahmanical body. It is vulnerable to touch by almost everybody except the brahman himself, provided he is not in an impure state. The brahman body must bear visual marks of its social position, just as the dalit body, especially "lower" castes of the dalits must bear visual marks of their social position. The brahmanical body is also fully mythological-allegorical, in the sense that each part of it has its own meaning, which in turn is made into a representation of the cosmic order of things. This is seen in the simple understanding of the notion of *kundalini*,[6] for example. An image of the body politic is available in the *purusha-sukta*, which tells of the origin of the four *varnas*. This is fairly well known, as is B. R. Ambedkar's sociotextual analysis in *Who Were the Shudras?*[7] of various other related myths of the origin of the varnas. As an aside, we could note the looseness of this body politic as compared to the Western, especially Renaissance, body politic, which is much more detailed in terms of the allegory (this is seen, among other sources, in Menenius's speech in Shakespeare's *Coriolanus*). Our interest here is not in the possible textual inconsistencies in orthodox explanations of the origin of the varnas but in the image of the body as seen in these more or less popular texts. Specifically, we are interested in the rather extreme vulnerability of the brahmanical body, which seems to be a measure of the vulnerability also of the orthodox social order. The purity of the brahman body is to be maintained at any cost whatever, reinforced each day of a brahman life, it seems to us, precisely because its superiority is tenuous. The extreme vulnerability is evident in daily ritual and other customs: You cannot put back into the plate a half-bitten morsel; you cannot be touched by others when you are in certain states; your own body can be a source of contamination. An aspect of the community is revealed here, which is that of the tenuousness of the affect of community itself. The more vulnerable a community is, the more likely it is that there will be frequent rituals. Inasmuch as rituals express communality, in some form or other, as performative expressions, or as possible dramatizations, or as plain unity, the frequency is more a mark of anxiety than anything else. It is to be noted that there is a marked difference in the periodicity of ritual across the castes, or, for that matter, across the varnas. There seem to be very few rituals among the dalits that run "the earth's diurnal course," whereas the brahman must perform certain rituals every time he wakes up, or, for that matter, visits the toilet. Therefore, it becomes necessary to distinguish among communities on the basis of the frequency of ritual. There are rituals in which the complete society must participate to some degree or other; these are usually annual and only rarely less than half-yearly.

The alienation from the body, a precondition for spirituality of any kind whatever, is actually given a social representation in these rituals and customs (inasmuch as practices could be said to be representations of some kind). It is only after this alienation from the body that any study of the inherited knowledge is possible. The alienation from one's own body begins, contradictorily perhaps, from the initiation rituals. These rituals, in their various forms, have been seen as rituals of entry into the community. Our observations about the alienation of the body show that this entry into a community is dependent on a negation of the body, a certain death of the body, a certain death of the centering of the self in the body. Society now takes over the bildungsroman of the body which we had left off precisely at this stage at the end of chapter 1.

In this ritual, which is also known as the second birth (thus all those who have undergone this ritual are called "twice-born" [dvija]), it is the body that must die for the second birth as a brahman to be possible. This aspect reveals a tension-ridden internal contradiction within the intellectual justifications of brahmanical ritual, which is based on a fundamental denial of the body: The body must undergo a serious and near-total regimentation and ritualization—it must be tamed—for the "soul" to come into existence. Thus, brahmanical philosophy represents possibly the most serious denial of the body. However, the body as such cannot really be mastered, even by denial, completely, ever. It is this inability to master the body in its materiality that generates the desperate intensity by which brahmanical ritual must affirm the superiority of the soul, along with the superiority of the whole caste, for what is brahmanical philosophy without the concept of the soul? The ritualization of the body also tells the person that the body no longer belongs to the person, since ritual is fully a social activity. The brahman in his house, alone within the circle of sanctity, performing his daily ritual is therefore a contradiction that is sought to be solved by alienating the body: The brahman is hardly alone; it is his body that is now alone, so to say, caged in rituals, manipulated, symbolized, allegorized, regimented, disciplined. A variety of movements and actions and operations are performed on it (by himself and by others) to ensure that sociality and community are maintained. The extreme form of this alienation of the body and its contradictory singularity and loneliness is hatha yoga; in Christianity the possible equivalent is self-flagellation. Thus, it is that a variety of hardships have to be undergone because unless these things are done (fasting, not drinking or eating anything while traveling, not eating food which is not cooked by a brahman—and that too in a certain way), the "world collapses." The maintenance of the world is the responsibility given unto you from the moment of initiation.

(It would possibly be disadvantageous to our argument that at least one strand of this philosophy flatly and roundly denies the existence of soul, but it is precisely this philosophy that has been, in a certain sense, "lost" [we have lokayata/ Charvaka philosophy in mind], its implications too radical for the upholders of "vedik" metaphysic and ritual. We return to this strand of philosophy at a later stage.)

Trần Đức Thảo's essay "The Dialectic of Human Societies" in his *Phenomenology and Dialectical Materialism*[8] explains some of the implications of such rituals, although these are drawn out in the book in a manner different from ours. Our description of the negation of the body is related to Thao's description and analysis of ritual. However, we do not see a successful dialectical operation here: The tension between the contradictory terms remains, and it must be tamed and controlled and exorcised every day, if not every moment of a brahman's life. That tension makes the contradiction repeatable every day, thus forcing a daily ritual.

Naturally, it is possible to think of brahman "becoming" precisely as the sublation of the contradiction, but the diurnal course of the ritual prevents us from suggesting that there is a resolution of, or relief from, the contradiction. The diurnal course is not necessarily maleficent for sublated becoming, but the intensity of the negation should make us think again. Moreover, even if the notion of becoming was admitted to be applicable to the daily and semantic moment by semantic moment of living, it is worth wondering why one would need to take up the same contradiction again and again,[9] if one had resolved it, and were relieved of it at a previous moment. It seems to us that the contradiction is maintained rather than resolved. Nevertheless, it is always possible to say that sublation itself is a continual process. We must reserve judgment on this.

It is not unimportant that the greatest threat is that of the literal touch of the dalit. This is rendered more or less impossible by the specific arrangement of regulations and social relations, but because of material bodily nature, it is not always possible to avoid metonymic contact, and most regulations operate on these metonymies, rather than on the literal touch itself. The world now is filled with possibilities of metonymic contact. It is not surprising therefore, that the sage meditating or performing penance alone in a deep forest represents one of the highest of brahmanical ideals. Somehow to be relieved of these possibilities of contact is possibly a factor in the ideal of renunciation as well, although renunciation is mostly encoded in metaphysical and philosophical terms that indicate that *any* kind of desire for things of this world is to be given up. This leads to the possibility that even these regulations on touch would be given up. It might seem as if metaphysics and philosophy are capable of transcending these mundane regulations. However, dalits do not have the right to undertake meditation in the forest; they cannot enter these realms themselves; if they do, they do so at the risk of death. Thus "society" is formed by various differences, but those who are different cannot escape society.

The regulations on touch operate mainly on metonymies, especially metonymies generated through possession. Since the literal touch of or by the dalit has been rendered impossible, there remains only the figural touch to be organized, and this organization must not extend the metonymies of contact beyond a certain limit. For example, it would not do to avoid breathing because there is a dalit nearby and some particles of his breath must be mixed with the air that is taken in by the brahman. Neither would it do to stop wearing something on the foot because a chambhar has made the footwear from scratch. Neither would it do not to allow a dalit midwife to touch a woman giving birth, especially in times of crisis. Therefore, we need to observe, first, what is thought of as metonymy and what is not, and second, which kinds of human action are thought to be contaminating, and third, what substance it is that has the power to be contaminated. We begin with the last first.

At this point we need to wonder whether there would indeed be occasions of touch between the dalit and the non-dalit, and we will have to contend that historically speaking, there is no occasion indeed when these would have to touch each other. We will have to give up the prevalent myth of a single "Hindu" society divided by castes, and we will have to suppose that several societies function with their own social organizations perceived as autonomous. The principle of inheritance ensures the continuity of this perception and cultural supposition. The supposition of autonomy entails the lack of superiority of this or that society over another, since difference also entails, in this manner of thinking, the lack of relationship. The economic relations are more in terms of necessary evil, or somewhat utilitarian agreements, which do not disturb the supposition of autonomy.

The one substance that cannot ever be contaminated is, paradoxically, the dalit body.[10] Being the agent of contamination, it cannot itself be contaminated by something else. It does not have the power to be contaminated. In contrast, with increasing gradations, the non-dalit bodies have the power to be contaminated and thus must *fear* the contact with dalit bodies. The power to be contaminated is stored up maximally in the brahman body—to be more precise, we should say the brahmanized body, since the uninitiated brahman child is not yet a brahman, neither are brahman-born women brahmans themselves in the same sense in which men are brahmans. We have characterized this complex ability with the somewhat paradoxical name of power-to-be-contaminated not only to accommodate the extreme vulnerability of the brahman body but also to emphasize the paradoxical passivity of the brahman notion of purity.

Some consideration of this issue is necessary. We have used the word *power* in relation with the word *contamination*. We are interested in pointing out not only the rather fragile nature of brahman hegemony, but also the rather fragile nature of power itself, taking into account both the positive and negative usages of the word *power*. It is almost self-evident that the brahman exercises his power through the notion of contamination—as a person, an individual, and as a social group bound by kinship and other systems of relation. What interests us at this point is the apparently paradoxical business of the relationship between contamination (vulnerability) and power (invulnerability). In the society under discussion, power is stored in the contaminability, if we are forgiven this crude English word formation. The more you are contaminable, the more power you will have (and the sociological notion of status is not inapposite here, because this has much less to do with economics than either capitalists or Marxists think. To be fair, we might say that status is economic power by other means and by that fact *more* powerful than economic power per se, since it can dispose of economic power with one spiritual shrug of shoulders, in a nonmonetary rhetoric of power, so to say). Our interest here is in the power itself (in its positive-creative and negative-coercive senses). At the level of application, it is the notion of the empirical touch of the dalit. At the level of exercise, it is the matter of the mutual interpretation and accommodation of rules and reality. At the level of form, it is that of status; at the level of substance, it is a matter of soul and body, of *atman* and *sharir*, or *kaya*. Our emphasis on the near-behavioral aspects of the empirical touch obtains a different texture and consistency at this point, since power must apply itself through specific, literal, empirical phenomena and actions, through the work done by apparatuses; and it is there that resistance too must apply itself. Resistance at this level is not a matter of attacking and touching symbols of power and self-determination (temple-entry), nor of kinship systems (Ambedkar's recommendation of intercaste marriages), but of inhabiting a social sphere where human beings can touch each other (in situations requiring touch, not just any and every odd situation) without consideration of power, status, or any other word or concept used to gloss over social inequality.

We have already seen that it is when the agency of touch is with someone else that it becomes maximally contaminating. A subterranean residue of this is to be found today as well, in societies that have forcibly unified over a long period of history. A dalit might think of the brahman as too distant to have any relationship, including that of cultural or economic exploitation or domination. "Their ways are different from ours": This perception is sufficient to ensure lack of any real (by which we mean mutually transformative) contact between communities. If the brahman avoids contact on the street, that is his style of life; it is not exactly as if we are seeking contact with him anyway. The point we are making is that it is mainly within a supposition of unity that a hierarchy is posited; it is mainly within this supposition, that Hindu society is held together by something (usually brahmanical ideology), that the question of contact and its refusal and denial arises. It seems undeniable that this supposition is fully ideological and needs to be given up. It should not be forgotten that the brahmanical description of the genesis of the varna system is merely a brahmanical description, no doubt others would have their own origin myths.

Among the various metonymies, those generated by possession are possibly the strongest. The very notion of possession is, as discussed in chapter 1, related to touch, in the sense that those things are ours which we touch most freely, handle, use; it is such things that we make ours. This is most clearly visible in the patriarchal possession of women, in marriage, or in bondage. The possessive case seems to be dependent on contiguity in a fundamental manner. Moreover, several regulations that are legal in form, in the sense that not following them generates punishment, are also related to these metonymies of possession. For example, the notion of theft as something wrong and deserving of punishment is fully and fundamentally dependent on possession. No one can steal from us something that is not ours.

The possessed thing is charged not only with these affects and figures, but also interestingly, with something of the nature of a usable thing as well. The possessed thing bears something of myself; something of myself is left on the things that we touch. In using a thing that belongs to me, we are *using, realizing, actualizing* a part of our own potential. Someone who takes our things away takes away something of oneself, some of one's own potential behavior. Similarly, when we give something that we possess, we give away part of ourselves. It is these potential forms of behavior that infuse the thing with value. The question of time is related to this, since the thing represents for us some form of our future action. From this point of view, it is easier to see that in exchanges of things, some potential forms of behavior and thus some versions of time itself are exchanged. The possessed thing also bears some encrustations of memory of its earlier use. It becomes clear from these observations that in exchanging things, one is also ensuring future forms of behavior: A gift is at once an expectation and an assurance of certain actions, and a confirmation of what we might call an "actantial" relationship, to use a word from A. J. Greimas.[11] While talking about gifts, their relation to these matters of action and practice must be taken into account. A study of such "gift-exchange" should take care that the notion of commodity does not surreptitiously operate in the description. The commodity-thing distributes time and action in a clear-cut way: At the moment of exchange, the labor that produced the commodity always belongs to the past, the consumption always to the future, and it is clear that unlike gift-exchanges, there is no question of exchanging and assurance of potential forms of behavior, but of finding equivalent amounts of labor—though we must also remember that there always is attendant the notion of use. Thus, although the exchange of commodity-things ensures *instantaneous* equality of exchange-value, gift-exchanges seem to function in a different temporality. It is too bold of us to do so, but we could entertain the possibility of conceptualizing what could be called "delayed/postponed use-value."[12]

There are also degrees to the affect of possession attached to things: The "thing" that is most charged with affects is the body. The curious sense of violation when someone unwanted enters into my house, for example, is entirely generated by this operation of metonymy.

Thus, the regulations of touch must first operate on these metonymic things. One's sense of territory is also the measure of how far these metonymies spread around oneself. Moreover, the things that we possess are charged with value by the fact of possession as well. Eventually, value may separate itself from the thing. We suggest that value (as related to exchange, which later will be encoded as economic value) first expresses itself in possessions, and possession is largely governed by what we can touch and how many times, at what expenditure of energy and so on. One can touch anything one wants; it is also a question of how much energy one has to spend to touch a particular thing. It is clear that something that is someone else's possession cannot be touched without some extra expenditure of energy.

We have seen how the body is the immediate domain of operation of social regulations on touch, and in the preceding paragraphs we have attempted to show how these regulations transfer themselves onto things. We have also stated that the major domain of operation of these regulations is the domain of metonymies. If one keeps in mind the earlier discussion of the difference between literal and figural, several interesting implications emerge. It is we ourselves, from our point of view, who have identified the figure; it is not necessarily true for those who possess the thing. Thus, the dalit's shadow is for us a metonymy, but it is not necessarily a metonymy for the orthodox practicing brahman. For him the touch of the shadow *is* the touch of the dalit. The same would be true for all other metonymies as well. That which is figural is turned into literal.[13] In this sense, the metonymic nature of these metonymies is nonexistent for the brahman.

What is true of the realm of metonymies should be true of the realms of other figures such as metaphor and allegory; what is crucial is the distinction between literal and figural. We now attempt to approach the question of occasions of touch from yet another angle.

THE SOCIAL OCCASIONS OF TOUCH: MAINLY IN THE CITY

Before talking about occasions of touch, we have spent some time in attempting a description of the general background on which these occasions become visible. *Visibility* is the key word here, because at some time in the nineteenth century, caste relationships become suddenly visible through the filter, as it were, of "injustice," *as* unjust. Needless to say, this is a filter that is made possible by some kind of modernity, encoded in acquisitive terms. The confusion between modernity and capitalism, in its rather advanced stages, is something that we will not attempt to clarify. This does not mean that there are no struggles in our world that seek to conceive of modernity in terms other than those laid down by the expansion of a mode of production that is, in its original form, regional and local, merely European. We do not have sufficient evidence, we think, to fuse together mixed genres with definite stylistic tendencies and alternative modernities. (This is one more reason why we have preferred the word *acquisition,* and its various polyptota and synonyms. The use of the word *acquisition* is meant to obviate a discussion of *alternative modernities.*)

Now that we have spent some time in attempting to articulate events in the nineteenth century as they become relevant to my concerns, it is proper to discuss the twentieth century and beyond, the epitomic present. By 1901, it could be said, capitalism, as we in India know it now, had taken hold. A jump brings us to the present moment, and the organization of touch is a little different, under the banner of modernity.

We have already pointed out in chapter 3 that it would be misleading to take the crowded bus or the crowded local train as a representative emancipatory experience of what modernity does to traditional forms of organization (in the sense that it does not matter who is touching whom). The organization survives, possibly in a modified, or a hidden, secretive manner. Our turning off of the sense of touch is rather secret, and everyone else whom we cannot but touch in the local has the same secret, and we know it and they know it, but it still remains a secret, because it will rarely be articulated publicly, especially not in the local train. It is a rather odd kind of secret, different from "open" secrets that everyone knows and gossips about. The community in the local train is a secret community, as far as organization of touch is concerned. We suspect that this is one of the secrets of modernity.

This temporary communality (or commonality) in the lack of the sense of touch (the denial of it, to be more precise) is, in a certain sense, the very apotheosis of modernity. The image of the city crowd, every member essentially mutually unknown (even if we are traveling on the same train at the same time in the same compartment and know faces by sight) is possibly a representative image. Some curious mutations of the sense of community develop in this space: Some festivals are celebrated; things are sold and bought; business is transacted.

Communities can be momentary as well, just as solidarities and affinities, and filiations and affiliations can be momentary. Social scientists have tended to look at forms of community that are strictly nonephemeral, but the crowd in the local train is a temporary—strictly speaking, ephemeral—community. (The relative neglect by social scientists of such ephemeral things is possibly an issue worth discussing: We might want to remember the business of structure and event, and the general tendency among intellectuals to prefer things of permanence rather than things ephemeral. Sociology must deal with rather stable formations such as the family, the church, or the state. There is almost no sociology of momentary things. It is not without significance that the word *momentary* is often accompanied by a general and indeterminate feeling of sadness and loss.)

Along with a sociology of momentary things, we also are in need of a sociology of nonobvious transgressions, as it were. For example, as Ram Bapat[14] pointed out in a conversation, there were in the past, several spaces where transgressions of established norms of touch were permitted. Secret societies were one such place—*tantrik* practices where norms of monogamous reproduction were neutralized into sexual activity with partners of other castes was not merely permitted but was the recommended norm.[15]

Nevertheless, there were disallowed sexual relationships taking place and with embarrassing reproductive results occasionally. There must have been a fair number of such children, who would have to be accommodated somewhere in the caste hierarchy, thus reinforcing the reproduction of caste. These are the really transgressive relationships, and their results must be placed, in a certain sense, outside of clan-rules of alliance in wealth and marriage. Children born of such relationships might want to claim rights and inheritance, and they must be denied. A whole legal discourse must develop around the rights of inheritance around this issue. This is not an issue merely of the illegitimacy of the child, but also of its caste. A classificatory system must be developed, and norms formalized and established. Such systems of classification are available, for example, in the *Manusmruti*.

Inasmuch as sexual touch is another realm of regulations on touch, these considerations become important. There is the legal sexual relationship between husband and wife, which must produce children, or the social status of both husband and wife diminishes; and there is the realm of sexual pleasure. A man's virility is an issue not merely of his manliness, not just a question of patriarchal, male-centered values, but equally a question of his ability to continue the line of inheritance. Among brahmans, the rituals on the dead body must be performed by the dead man's son, or some legitimate male inheritor, without which the release from the mundane becomes difficult. Therefore, it could be said that inheritance is a greater value here than the maintenance of male domination. If a man is incapable of reproduction, he loses quite a lot of the social privileges that he would otherwise have enjoyed. The principle of inheritance often uses both genders as instruments through which it both hides and reveals itself.

The strength of the principle of inheritance becomes clear in this organization, which is an organization of touch in a somewhat distant manner, but at the same time it is true that both the principle of inheritance and the organization of touch are mutually supportive. Having indicated that there were forms of behavior available in the pre-eighth-century world in which men and women of different castes could come together for sexual purposes, we now begin to consider the nineteenth century.

If in the first half of the nineteenth century, the major institution in which different castes had to come together under the banner of economic achievement and acquisition, was the Native Army, then in the second half they had to come together under the banner of a much more economically oriented form, that of material commodity production itself, and instead of the military camp, the factory and the mill and the labor camp became the loci of such a congregation of castes. We are talking about a metropolis like Bombay.

In their book of 1889 called *Mumbaicha Vruttant* (An account of/ report on Mumbai),[16] Balkrishna Bapu Acharya and Moro Vinayak Shingane give us some invaluable details of life and times in Mumbai of the period. They report that in 1881 there were 20 hotels, 23 refreshment rooms, 49 taverns, 174 retail liquor shops, 216 *tadi* (local palm liquor shops), 10 boarding houses, 8 lodging houses, 4 ice-cream saloons, 36 coffee-houses, 9 chandol-khana (opium-houses where you could also smoke and/or consume other substances). They also state that there were 51 shops of various arms and ammunition, whereas at the time of writing, which is about seven years later, the number of retail liquor shops had gone up to 159 Western liquor, 347 country liquor, and 100 "*tadi*" shops.

The total number of people in Bombay according to the 1881 census, on which Acharya and Vinayak often depend, is 773,196. Out of these 35,428 are brahman, 407,717 are "Hindus of other castes," and 40,122 "lower caste Hindus." The density of population is worth noting. Whereas in Calcutta there were 208 persons per acre, in Bombay, there were 759 persons per acre, yielding about 5.1 families per house/ building. There are 27 places from which you can hire a vehicle (usually drawn by horses), and there are 14 representatives of other nations and/or kingdoms. The details on business and industry are also useful: Under the heading of famous "Native Companies," the authors list 19; under famous "European Companies" there were 53, some of which are still operative, like Fevre Leuba, Lund and Blokeley, Phillips, Richardson and Cruddas, Greaves (then Greaves Cotton). Under the heading of banks we have a list of 10 banks. Under "Mercantile Associations," there are 9 worth listing here: Bombay Branch of Fire Insurance, Chamber of Commerce, Cotton Traders' Association, Mill Owners' Association, Native Merchants' Association, Traders' Association, Underwriters' Association, Bombay Bible Women's Association, and Bombay National Muhammadan Association.

There were 56 spinning and weaving mills, 11 flour and oil mills, and at least 4 photographers, 28 libraries, 3 dailies related to business: the *Daily Merchants' Companion*, the *Daily Commercial Sales Report*, and the *Chamber of Commerce Daily Trade Returns*. There seem to be 14 clubs, including the Rippon Club, the Canara Club, the Brahman Club, and the Saraswat I, the Arya Brahma Samsad.

The above list is meant to give a snapshot of possible social and economic interactions, and it is not indicative of the totality of newspapers, commercial houses, and other activities in the Bombay of the period. The rise in the number of liquor shops, for example, does tell something, as does the number of weaving and textile mills. It is in these circumstances that we have to see the various configurations of regulations on touch. It is well known that the major business in Bombay was that of textiles. It is here that we will be able to see that already a number of people have given up their inherited professions and come to Bombay in the hope of making money. It should be possible to argue that the economic arrangement itself makes it impossible to retain the inherited profession, which now is paying less than what it would have in the earlier periods. The disturbance of the earlier set of relationships of exchange now forces people to give up their inherited professions and hope to acquire more benefits, culturally and economically, by adopting new ones. Nevertheless, in this period too the associations of caste continue to operate. In fact, a number of them operate till this day. Thus, in the spinning and weaving mills, for threading the bobbin, you have to put the thread into your mouth, wetting it to make it easier to insert. It is well known that dalits were not really allowed to work in this section of the mill, since this is not just a question of an accidental touch of the dalit as you walk out of the door, but something more dangerous.

Sometime in the beginning of the nineteenth century, the serious mixture of genres of inheritance and acquisition begins to take place. The first three decades of the nineteenth century are really crucial from this point of view. It is in these decades that caste issues had become important in the Native Army, as Seema Alavi has indicated. The establishment of the railway and the telegraph in the 1850s are possibly the turning points, since they changed the style and efficiency of administration. In Maharashtra, the defeat of the Peshwas in 1818 is an earlier turning point. By 1857, it became possible for a chambhar to ask a brahman for a jug at the riverside. That story is emblematic, and represents the changes in the relationships, and especially the thoughts of dalits. These thoughts continue, in some form or other, for a very long time.

What does it mean to say that there are regulations on touch? This is not, as we have pointed out, a matter only of touching the bodies of this or that type of person. This is a matter of the organization of social relations among persons and communities. These regulations operate at all levels of the organization. The most blatant examples of these regulations take the emblematic form of being given a separate set of things to use in a restaurant, for example: There are earthenware cups that can be easily destroyed after use; you are not given the kind of cups that others use. You are possibly not even allowed inside the restaurant and must sit somewhere near the steps to have your tea, if you have the money. Already the question of appearance becomes extremely important: If you are wearing Western dress, without any caste marks on your face, then unless you tell them your caste, they would not know. They cannot trust you to tell, so they will perforce ask, and treat you according to your station in life. The disappearance of caste marks, especially from among the dalits is a matter of concern, and caste, shorn off its visual signs, must now be identified by another visual classification: face type, skin color, general physical build, which come to be polarized into "fair brahman face" and "dark dalit face." The already established discourse of race comes in handy here. The question of the origin of the Aryan race, the discussion and debate about the "Arctic Home of the Vedas," serves two purposes: one to align and attempt to equalize the brahman and a few others with the Europeans, and the other to push the dalit into the lower levels of social organization, for it is assumed that the Aryans are superior.

What is crucial for our understanding of the social organization of touch is this new space in which a variety of people are mixing together, without any clearly defined caste marks on them. The persistence among brahmans of these caste marks must be viewed in this light. The sacred thread, the mark on the forehead (and its types as well), in short, the whole optical semiotic of caste appearance, so to say, must now undergo a change. (In this context, the revival of the gandha[17] by political factions whose major constituents are nonbrahman "upper castes" is, from this point of view, an interesting shift in the semiotic.) The disappearance of clearly visible caste marks among the dalits must generate a great anxiety about touching a stranger, and the established code can become only more rigid: best not to touch any stranger because who knows what his or her caste is. The neutralization of touch with which we started the discussion has these antecedents. It seems to be a remarkable phenomenon that around the same time that caste becomes "visible" under the heading of "injustice," the empirical visibility of caste marks is beginning to wane.

Our hunch is that the sense of touch that must be neutralized in the bus and the local train, but which must be charging up the body with great amounts of libidinal energy—after all it is a literal touch—must find release, and it finds release in violence, the general sexualization of commodities, and the discourse of hygiene. In upper-caste ideology at the present moment, caste is understood in terms of hygiene and cleanliness (these two are not distinguished from each other); we must not touch the dalits not because traditionally they are untouchable but because they are unclean, their language is impure, and so on. Several complementary discourses come together at this node and must be separated by driving wedges between them. First of all, it is clear that hygiene, health, and cleanliness are not really dependent on one another: otherwise, all these "unclean" people would not survive for us to find them untouchable; their own uncleanliness and unhygienic lifestyles would have spread epidemics of diseases and killed them off anyway. That the people who live in these unclean slums live on, is proof enough to separate health and hygiene and cleanliness. We need to drive the wedge a little deeper, by questioning the very notion of a healthy body as a moral ideal. The discourse of medicine, the discourse of sanitation (often reduced to soap and detergent advertisements on television and in other media), and the discourse of caste come together in this and remain coherent in this ideology even if there are contradictions between them. That health is a moral ideal, and a long and healthy life is a moral ideal can be seen quite clearly in society. This valuation of life is rather different from earlier valuations to be found, for example, in Epicurean themes of having fun and seizing the day. It seems to us that the ideal of a long life is held up as a moral ideal as well for the very simple expedient that it encourages consumption and commodity exchange: The longer we live, the more we will consume, the more we will continue to buy things. We need to take care of the body, not because it is a source of pleasure or knowledge, but because it is in itself good to live a long life. The self-evident moral good assumed in the nature of the assertion of the ideal of long life must be questioned, not merely because it is an ideal, but because this ideal is deployed and manipulated for certain benefits. Far more than the discourse of health, it is that of hygiene and cleanliness which assists in justifying caste divisions and untouchability. Thus, the inherited profession of cleaning up other people's mess falls upon those castes who have in any case been doing that for eternity—it still *is* their job to collect the city's garbage.

One of the fundamental features of a city is that there are more specialized professions in a city than in a village. There is a firm or a company or a specialist for everything, and professions are often internally divided further. In a village the carpenter and his family would undertake repairs of cartwheels, making doors and windows, and all such objects; the ironsmith would undertake most activities related to iron and so on. The whole domain of production and maintenance belongs to such a division of artisanal skills. In a city, these would get divided rather differently.[18] The skills of maintenance would be valued less than those of production, since in any case, objects are ideally made in two ways: Either they should not need to be maintained, or they should be scrapped, and new objects purchased.

The greater differentiation of professions fragments the visibility of caste communities, which recede from the sphere of work and profession to the more domestic aspects and the more cultural aspects such as festivals and marriages. The home from now on becomes the major sphere of caste-based activities. This is generally true of several other activities as well, including religion per se: Given the inability to practice one's religion in the domain of the public in general (where one also has to accept the legitimacy and right of other religions to practice and fashion themselves in that sphere), the home becomes the location of religious practice. (Clearly, this is indicative of a certain "intolerance": It is because one does not want other religions to be practiced "publicly" that one practices one's religion at "home.")

We notice thus the two directions in which the regulations shift: One is the direction of more elasticity and adaptability; the other is in the direction of rigidification, but the location of the more rigid practices has shifted from the street to the home. Other developments also contribute to these shifts. In urban centers, the very organization of family housing determines that there be no traditional communal relations among the occupants. Houses are organized rather like news in newspaper columns; there is no relationship between the news in one column to the news in the one to its right, or the advertisement to its left. The organization becomes more abstract, in the sense that it is not immediately manifest and needs to be explained by taking recourse to some principle that is transcendent to the phenomena.

Something like an inversion takes place in these shifts. Practices that would be rigid in the public become elastic, and those that would be elastic become rigid. As is clear enough, in the secret societies, secret sexual liaisons and exploitations were the areas in which the regulations were a little flexible, but in public they had to be rigid and firmly practiced. The relief from regulations is now available in the public domain, but not in the privacy of the home, where earlier one could sexually or otherwise exploit, say, a dalit woman or man.

There are several changes, and these are mostly of the form a further rigidification, which results in violence. Faced with emancipatory agencies and emancipatory state policies, the principle of inheritance is sought to be asserted by violence. From this point of view, the most serious and bloody battle between societies of inheritance and societies of acquisition is being fought in nonurbanized areas. It must be remembered here that hereditary possession of land is the major issue of contention, from which follow the rest of the issues that could be deemed to be more cultural. In a remarkable way, it is the state that has to, willy-nilly, and in its own "political" way, take up the responsibility of emancipation, through what is otherwise thought of as a coercive instrument, the police. This instrument, along with the other instruments of the judiciary, is susceptible to manipulation by those who run the government and the opposition to it. This, without doubt, is the most obviously "political" aspect of the anti-caste struggle, which now must disguise itself as a human rights issue, which perforce must seem less susceptible to political manipulation than a nonambiguous anti-caste struggle (which itself has proved quite susceptible to such manipulation).

The struggle between societies of inheritance and societies of acquisition is also the struggle between tradition and modernity, with their special cultural encoding at present.[19]

In urban centers, the substance on which the regulations operate is also undergoing a shift. Inasmuch as there is the situation of people of different castes forcibly mixing together under the rubric of acquisition, the body begins to stray out of the orbit of regulations. The regulations must now operate on some other substance. In most cases, this would mean only a further strengthening of the regulations on other substances such as connubiality and commensality; in some cases, this would lead to stronger adherence to religious principles. However, in the restaurants and the trains, the ostensibly public places, the bodies themselves are more or less without regulations of the traditional kind. The neutralization of touch discussed earlier is a new kind of regulation on touch. Without doubt, the regime of regulations on touch differs from place to place and needs systematic and focused study of a kind that we cannot undertake here. It is also to be noted that the meaning of the gaze also changes from place to place; different regimes of looking and seeing and watching operate depending on the place.

Thus, the most prominent occasion of involuntary touch is to be found in these public places. In cities, this has already transformed itself into an occasion of looking rather than touching. The regimes of gazing and looking are different in cities and in rural areas because of this. For the most part, gazing comes to substitute for touching. One does not, if one is a member of the urban middle class, usually gaze upon the drunk lying in the street or the beggar with his or her "deformed body."

The more communal kind of touch is still obtained on festive and ritual occasions, where the inherited meanings of touch still operate across generations. There would be only the minimal regulations on these occasions, since the members of the community may, in any case, touch each other more freely than others. These occasions are comparatively less frequent and do not need a detailed consideration.

At the general level, then, neutralized touch is released in other domains, usually those of violent touch. This necessarily operates across communities. At the level of the individual, we have pointed out how neutralized touch is released in the domain of sexuality, hygiene, and caring for one's body. We have also pointed out that in general the regulations on touch in urban and rural areas shift in different directions and are used for different purposes.

Finally, we need to understand the question of modernity and tradition. Different notions of time operate in societies of inheritance and societies of acquisition. In the former, the present is a detour that the past takes in order to arrive at the future (the future is rather like the past), whereas in the latter, the future is radically or minimally different from the past and the present. This makes for the modern concern with "time in its open-endedness." All movements for social change must necessarily presuppose that time is open-ended, that "anything" can happen, if only we are willing and able to make it happen. Fundamentally, the source of value (and we must note here its relation with time) is human in the modern conception (whether capitalist, Marxist, feminist, dalit, or some other conception). From this point of view, societies of inheritance could also be called societies of repetition or iteration and citation, whereas societies of acquisition turn out to be those in which something new, something that is not a mere citation, can be said. No doubt there will be present the questions of citing what in which context, repeating what in which situation. Neither should we forget the role that memory plays in societies of inheritance: Citations and repetitions are fundamentally dependent on memory (otherwise every citation will be perceived as something new), and as we have already pointed out, the role of memory in such societies is very important.

It is worth noting that in the history of the anti-caste struggle, it was only after a certain body of writing and protest had been built up that people began to speak of the "long history" of the oppression of the dalit. This is a curious phenomenon, in a certain sense, since it is precisely in a society of acquisition that these apparently primordial memories are being jogged.[20] The past is cited now in a much-changed context. The struggle between those who wish to preserve the caste system and those who wish to undo it is also a struggle between two differing notions of what the context is. We have already indicated quite clearly that the discourse of the anti-caste struggle is fully and fundamentally committed to a notion of time in which the future *must be* different from the past and the present.

THE OCCASIONS OF TOUCH: TRADITION AND MODERNITY

In chapter 3 we discussed the nodular form of organization found in the cities. There is something interesting in this nodularization of work space (Fordism and Taylorism, Post-Fordism), and living space (individual flats in apartment blocks) that results in a segmentation of society. Such a nodular organization is present in rural areas too, but in a vertical manner, so to say. In societies of inheritance, the nodules are vertically organized to form unmixing stacks of clans and caste groups. In the city, in societies of acquisition, the vertical arrangement is disturbed, because some other principle is now raised up to form the vertical axis, and this, we have already seen, is the principle of acquisition. However, precisely because there was a kind of nodular organization present, this change of principle in the vertical axis is either overvalued or undervalued. In societies of inheritance the principle along which things were raised up or pushed down on the vertical axis was that of inheritance itself. The disinherited, those without any inheritance, did not have a high place. We are suggesting that this organization is nodular because it seems to us that in principle this prevents any intimate contact between the dalit and the brahman and other high castes.

In modern urban centers, the principle on which people and clan groups are raised up or pushed down is that of economic acquisition, and next to it, other forms of acquisition. This also creates its own nodular organization, in which the artist rarely comes into contact with the industrial or mana-gerial executive (unless the artist is looking for sponsors). But this very nod-ularity of organization is what the principle of inheritance uses to sustain itself, by changing as little as possible. Thus, areas develop in the city that are dominated by certain castes, certain religions, just as the city itself gets segmented into workplaces and residential places, shopping and parking ar-eas. These are rather different from the ostensibly caste grouping of houses in societies of inheritance. Quite often clan and caste loyalties and kinships might even be further strengthened in cities because of the greater pressure for economic survival in cities. We have seen how the transmogrification of the people outside the boundaries of communities is an internal neces-sity of communities. This transmogrification also operates a little differently in the urban areas and quite often works mainly along economic lines: Those who are not of the same economic status as we are become "the others" (of-ten transmogrified into what Heidegger calls "the they" [*das Man*]), to be hated and loved, depending on whether they are higher or lower than us. But even here, when it comes to touching, and commensality and connubi-ality, the principle of inheritance still operates underneath the lacquered finish of economic interests and, therefore, rational interests. In fact, if the lacquer is thick and does not allow the inner substance to breathe, the lac-quer develops cracks to reveal the inner community. The segmentation of workplaces and living spaces does not allow the inherited caste community to continue in the same form as it had in societies of inheritance. Never-theless, an imagined form of community still operates, and individuals find sustenance in it. Religious communities now are of this kind. They are al-most virtual communities in cities that are seriously segmented and nodu-larized. The confused styles of architecture, art, and other cultural activities are witness to the segmentation and nodularization of communities. In the world of art, certain segmented forms of appreciation and patronization develop. The viewing of films gets segmented into "commercial" and "art" cinema, with different cinema halls and often different patrons. In fact, segmentation of the kind that we are describing is most clearly seen in the arts. It would not be incorrect to say that the market itself is segmented, with certain shops catering to people with money, others to people with less money, and still others to people with very little money. Members of these segmented communities rarely relate to each other. Needless to say, these communities, with their surface unity under the heading of economic interest and acquisition, are extremely fragile and momentary. The affects associated with a contract and with a family reunion are different and reveal the fragility of the former.

The overwhelming interest in economic survival in cities, that is to say in societies of acquisition, covers up our perception of caste, which now must change, adapt itself to the new conditions, and take new directions. The mixed style of society that we have conceptually posited is revealed most clearly in cities. Clearly, there are as few occasions of touch as in rural areas. Living in an apartment block, the question of touching someone other than the immediate family arises only when one travels in a bus or a local. In fact, it will be seen quite clearly that economic advancement further curtails the occasions of touching someone other than immediate family: If one travels in a car rather than a bus or a local train, then the question does not arise at all. Depending on the economic power, which now gets quite firmly associated with what we have called the power-to-be-contaminated, the frequency of the occasions of touch changes. For people living in small constricted spaces but in great numbers, the very sense of touch is operative only on sexual occasions. A whole group may lie down to sleep beside each other, several limbs strewn over bodies of others', but this may not be seen as touching at all. The neutralization we have discussed operates in such spaces as well. In this space, touch is not only neutralized, but quite often loses its meaning altogether. We would have to say that the only regulation that is operative is that of altruistic and sexual touch; all other meanings, including the division between touching oneself and touching others, are worn so thin that they almost disappear, under the circumstances described above.

In both societies of inheritance and societies of acquisition, one touches things more than persons. This is one more reason why regulations on touch operate more on touching things than on touching people. In societies of acquisition, the things one touches are mostly commodities. We have already pointed out the sexualization of commodities and their advertisement and consumption. The proud (somewhat middle-class) car owner wiping his car every morning and the woman of the house, if she is a housewife, equally proudly cleaning up the house are touching things that they possess with a love that can be found only in societies of acquisition. A whole new realm of touch is opened up within the segmented apartment block, within the family.

Now that we have concentrated on the family in the city, the differences between the family in the rural areas and in the city need to be discussed briefly. It is almost a cliché of modernity that in modern societies, there is a breakdown of the traditional extended family in India (what the official system called HUF: the Hindu Undivided Family). It seems to us that segmentation is a much better metaphor to describe the changes that have taken place in the family. It is not as if each husband and wife living separately are a family by themselves. Often enough, brothers, sisters and other relations come together under the patriarch's or his wife's umbrella, so to say, and settle family issues, usually of marriage and money. It is much more the case that the family members who used to live in a connected, contiguous space now stay in separate places and travel to meet each other. This does not amount to a breakdown, although the frequency of meeting would depend on the distance between the segments and the amount of time and money to cover that distance. It is within the segmented family in the city that we can see most of the features of the conflict between the two stylistic tendencies in societies, especially in second-generation migrants: The older generation must perforce stick to an inherited profession, whereas the second generation, born and brought up within the city has a choice of professions available to it. Education, far more easily available in urban areas than in rural areas, is already playing a role in the lives of the second generation. In any case, the city is, in modernity, the most privileged place, a place of modern pilgrimage.

Within the family itself, the discourses of sanitation and sexuality are dominant, second only to the discourse of money. There are several regulations on what children can touch or not touch, depending on the monetary value of the thing involved. In a traditional unsegmented family, the regulations on children are less severe, it seems to us, because the kind of surveillance of children that is present in urban, segmented families is not present.

It becomes increasingly clear, the more we think about the city, that quite a few of these changes are caused by the particular arrangement of space, which is always scarce. The surveillance on children increases because the children have to sit at home, and the parents can easily watch them. A different situation obtains in the houses of people who do not have to live in apartments. Most other observations that we might make of these phenomena are more or less obvious, and we will not reiterate them. It is time to return to the main theme of the discussion.

THE NOTION OF SOCIAL CHANGE AND ITS IMPLICATIONS

Imagining the future as radically different from the past and the present creates its own complications: If the future is a break from the present, what form of action will allow the present to reach the future, to bridge the gap of the radical difference? Theories of social change must have a clear-cut, well-defined, and solid bridge that would take one across to the future, but the future is posited on different and better principles, and the bridge remains too caught up on this side rather than methodically and methodologically connecting the present and the future. For a real and material passage, the future must be seen as a continuation of the present, but that is precisely what social change does not need. Imagining a different future and working toward it in the present involves one in a constant and interminable difference between the future and the present, which leads, logically speaking, to a paradox, or, if you so prefer, an aporia (the future, so different from the present, will be reached by working assiduously in the present).

All forms of social action and the ethics supporting them must remain caught in this impossibility. Moreover, there are competing futures that not only compete but struggle with one another. The future that is the favorite of societies of acquisition, especially those that produce commodities that need marketing and advertisements, is the future that is already here. Thus, the announcements of the latest technologies assure us that we are catching up with the future. This is also a question of the linear or nonlinear sequencing of memory, perception, and expectation, of the past, the present, and the future. The very notion of acquisition needs an interminable linear sequence. The principle of inheritance, often confused with the cyclical notion of time, is rather different: It seeks to re-create the past in an interminable race with the past itself.

The implications are significant, and they range from the framing of constitutions to policymaking for education and economic change, and the most significant implication is for activists who dedicate (and often lose) their lives to social change. The danger here is either of thinking of the future as an inevitable continuation of the present and ignoring the radical unknowability of it and in a blithe spirit of continuing only to do what is right from the present point of view or making the future so different as to not know how to reach it.

Given the segmentation (not to be confused with fragmentation) of societies and families, the notion of politics as a collective activity with its own formation of communities and its own rituals of commitment to a common ethic seems to be under threat. The segmentation is overcome, and people gather together only under the most immediate, and given the media, the most catastrophic "causes." Politics itself becomes nodular, and leaders become leaders of rather small factions, compared to the mass following of leaders of the earlier half of the twentieth century. Thus, when we think of politics before 1947, and politics in the '80s, '90s, and beyond, we are thinking of some rather different kinds of politics: The politics in a "segmented market society" must be different from the politics of a "yet to be fully formed market society" (where by "market society" we mean the Western model of capitalist "development").

We might wonder why this segmentation is a necessary phenomenon or, rather, who benefits by this segmentation of society. At the level of ideas, the notion of segmentation is different from that of hierarchy and rather more comfortable. At the level of actual relations at the practical social level of monetary and other economic transactions, the segmentation makes for better manageability, through making it easy to gather data on the participants. The comfort afforded by the notion of segmentation should not be underestimated: After all, we all have a conscience that needs to be soothed in some way, and the notion of segmented markets (that's not where we like to shop; that's not the kind of music we like; what Pierre Bourdieu used to call "distinction")[21] offers us this soothing balm. Difference in taste, acquisition, and tradition can now be understood in, and can be reduced to, terms of the kinds of things that "we like to buy" and the kinds of things that "they like to buy."

5

TOUCH AND TEXTS
Ancient and Modern

THE PLACE OF TEXTS IN SOCIETY AND ITS ANALYSIS: CLASSICAL TEXTS

It is here that we discuss the issue of classical texts. We are more than aware of the vast number of texts that are available, from the Vedas onward. However, the argument that follows does not depend on any knowledge of some of them, since it is an argument purely in principle. As is well known, great debate is possible on the supposed meanings of the differences between a tradition written down and a tradition oral. It is a brahmanical myth that "Hindu" society is a laudable example of an oral tradition. Three arguments can be made here, two of them complementary but different within the framework of the brahmanical myth and the third outside its framework. The first is simple: These texts, the various Vedic, Brahman, and Puranic texts, were always written down anyway. There is a tacit acceptance of the fallibility of human memory in the writing down.

The second is a little less simple. We assume for this purpose that the business of writing down these texts—all of them—is taken up in the first place in order to stabilize, standardize, and preserve them in the assumed purity of their supposed origins. The various techniques of memorizing that evolved within orthodox brahmanism for the preservation of oral texts, especially of the Vedas, ensure that the texts retain their stability. After mastering the text in a variety of combinations of letters, forward, backward, skipping, or exchanging every second or third letter, and so on, the text is as stable in memory as any text written down and preserved in that material form; it is as good as a written text, and the distinction and supposed difference between written and oral collapses. It can be referred to at any point in time and debate, and depending on cues, relevant passages cited. In fact, it is further possible to argue that the material on which the text is preserved in writing is more capable of corruption than this solemn, sacralized, and dutiful memory—it is more written than texts written on material surfaces. Within the essentialist-idealist framework, texts are strictly nonbiodegradable because of this specific use and operation of memory—and we have in mind here Derrida's essay on biodegradables.[1] What the techniques of memorizing indicate is the effort that was made to maintain the repeatability of the text, and, in a case of the tail wagging the dog, of maintaining a certain social hierarchy by being able to repeat a text. It is perfectly understandable, from this point of view, that education was equated to being able to repeat convenient pieces of nonbiodegradable texts in real situations of tension and conflict. Transgression of the existing social order is rendered almost impossible, since the memorized text as such (because it consists of such a variety of things) contains all manner of disparate elements, which is what we tried to indicate with the earlier adjective *convenient*. These disparate elements can be invoked in their purity and sanctity at moments of convenience, thus making it very difficult to transgress the existing social order.

How to ensure the transmission of this memorized, "identical" text is the next question, and here a principle that is already operative in society comes in handy: the principle of inheritance, which can, as with property, so with memory, easily be equated to that of male lineage. Such lineage need not always move down the line of father to eldest son; it can also move sideways (uncle-nephew, as in the case of Bilbo and Frodo Baggins in *The Lord of the Rings*), and, more important, it can also include disciples (as is sometimes the case with the transmission of the Hindusthani classical musical tradition). The incredible and cruel emphasis on memorizing in brahmanical education (the training for which began, in earlier times, at a very early age, the moment a child was initiated) needs to be understood in this light, and its residues in the modern world in the notion and in the practice of education and examination identified. It is necessary to wonder, for example, whether it is the initiation that grants a person the right to remember and memorize, the right, so to say, of memory, which is granted along with the right to learn to read and write.

As is well known, the dalits of the past have left behind almost no ex-plicit record of what they were, what they did, or what they said and what they felt; and we believe this is so because the function of memory in the consciousness of dalits (not to be confused with "dalit consciousness") is dif-ferent from the preservative function of memory among the brahmans. It is possible to argue that with the anxiety that memory will corrupt the past, it became necessary for brahmans to write it down, leave some record of one's knowledge of the past. With people who had nothing to invest in the past, and whose memories were not arranged along preservative lines, there was no felt need to leave records of the past or even to narrate the past. In any case, within living memory, if the narrative of the past is also a narrative of suffering,[2] there might be a tendency to forget it, rather than remember, memorize, memorialize, commemorate, and rememorate it. The issue of the nature and function of writing in societies is also connected to the different uses of memory found in the brahmanical and the dalit traditions.

Then, are dalits to be considered to be in a process of perpetual becom-ing? On the assumption that becoming can never catch up with itself to be able to remember itself and preserve a specific moment of becoming, in memory, or in some other form like writing, it would seem that dalits (and women, tribal people, and most illiterate people) might be understood to be in a continual, elusive process of becoming. It is through this that we might find an explanation of how women, dalits, most tribal people, and most illiterate people are thought to be closer to nature, and it is through this that we might be able to identify the socially real and philosophical pro-cedures that privilege writing and "culture" as such. The question that begs to be asked is, are dalits then without history?

It is to be remembered that women too are prevented from learning these things, whether in the simple sense of learning to read and write or memo-rizing sacred texts. Women do not have the right to initiation and, in princi-ple, could be said to be a different world altogether. Anthropologically speaking, inasmuch as quite a lot of the caste distinctions also operate in terms of marriage, and since marriage involves some control over women and not only their sexual behavior (not to be confused with women's sexu-ality as such) but also the economy of their labor and the possibility of in-heritance, the question of caste relates to gender as well. As is clear, the question of caste is made fundamentally to depend on the right to initia-tion (and we will discuss this later yet again).

The texts are written down. The power enjoyed by people who know how to write over people who do not need not be elaborated. What is equally interesting is that not everybody in the community is literate quite exactly, but the community functions as a literate community. Brian Stock gives an interesting description of what he calls "textual communities" in *The Implications of Literacy*,[3] in which he also discusses problems of interpretation. Stock makes the convincing point that even though a large section of people may not be able to read or write, a community is built around important texts, and ideas from the texts circulate among the people. Thus, a community forms around texts, but the community contains people who not only have not read the texts but also lack the skill to do so. Without doubt, the problems inherent in all circulation of ideas and texts are to be found in these communities: Ideas from the texts might get distorted, mythologized, quoted for a variety of purposes, or just be different from what is written in the pages of texts. Ideas in the text get translated, transformed, and used in senses and situations that might be incompatible to the text if we were to stick to the letter. Even if one does not have the skill to read the *Manusmruti*, and in addition may be disbelieving of what a brahman might say the text says, one may have some idea of what it says and want to abide by it. In this way, texts might even come to have a certain authority over people who do not know how to read, and ideas from the texts circulate among the people. (This is true, for example of the stories from the *Ramayana* and the *Mahabharata*.) Over a period of time, as might have happened in our context, these ideas may come to be *the only ideas that are in circulation* (and not merely be the rulers' ideas that are also the ruling ideas).[4]

The point we are attempting to make is that inasmuch as the brahmanical tradition is deeply and fundamentally a textual tradition, the operation of these texts within and without brahmanism must receive some attention. It seems that these are available also within the "oral" traditions of the dalits and might provide some interesting points of interaction between the written versions and the "traditional oral" versions. Perhaps equally important, the dissemination of brahmanical ideas of ethical behavior needs a similar study, provided that it is conducted along with a study of nonbrahmanical ethical ideas.

The question of the organization of, and regulations on, touch receives a stable determination from these textual practices, because when confronted with some anomaly or difficulty, it is the texts of the brahmanical tradition that would be consulted, and specific hermeneutical principles would be invoked to justify specific applications of the injunctions in these texts. These issues of interpretation necessitate a discussion of the relevance of the brahmanical texts to the analyses of caste and of society in general. It is abundantly clear that these are thought to be extremely important for understanding caste and society. This is not just a question of the enchanting philosophical profundity of these texts: The question is also whether we are going to be enchanted by the philosophical profundity of, or, moral outrage at, these texts. The case is a bit like ancient Sanskrit texts on Indian classical music, which, in the form in which they survive and are passed on and venerated, give absolutely no clue whatsoever as to how to produce the notes that are given the specific names they are given: Presumably the material of which the instruments were made was different; there were organic strings in use; the wood was different; the very sound of the note was different. The same is, to a large extent, true of texts on other subjects as well, and this fact throws up impossibly large questions about interpretation, which is to say, large questions of politics. That politics are a matter of rhetoric and interpretation is clear; that these are complementary notions is something we learn from what is, in its own context, a minor point in Hans-Georg Gadamer's *Reason in the Age of Science.*[5] In most traditions of rhetoric and interpretation, grammar (which is to say some form of structuralism) is an important issue. Here we understand grammar in terms of syntax, which we believe to be a matter of organization, tradition, institution, and institutionalization.

Allow us to give a minor and known example. The ritual of initiation we take to be a process of institutionalizing a child into an adult. Initiation not only institutes the child as an adult, but also introduces the child to the idea of rupture or near-total break. We are speaking mainly of the so-called dvija tradition. No doubt there are rituals of initiation in the dalit tradition as well, but it is not clear whether they perform the same function. The question of caste is known to turn on the right to initiation, the very ceremony that kills you and gives birth to you. We are forced, in a certain sense, to alternate between anthropology and grammar for the very simple reason that the brahmanical tradition points out that the dalits are different from the other three castes because they do not have the right to initiation, which right they lack because they are *not* mentioned in the text that prescribes—enjoins, rather—the initiation ceremony only to the brahman, the rajanya (kshatriya), and the vaishya. The text says that these should be initiated in the seasons of *vasanta, grishma,* and *sharada.* The "fourth" are *not* mentioned, and by the principle known within the tradition as the "power of the second case" (the accusative),[6] and known in European jurisprudence as the Latin legal principle of *expressio unius est exclusio alterius,*[7] they lose the right to initiation. We have discussed this elsewhere[8] and must refrain from repeating ourselves. This, as we said earlier, is the extreme form of the argument against the Sanskrit texts and the compulsive and immediate use of them the moment caste is mentioned: These create the impression that an unjust social practice is the result of rigorous grammatical interpretation and therefore demonstrate logical and metaphysical consistency. The picture that emerges is of a society in which these textual traditions serve as the nervous system of the social body, so to say. No doubt this is the picture that many want and, perhaps therefore, like.

The third argument, outside of the framework of the brahmanical myth of the "orality" of "Hindu" culture, focuses on the contradiction between what might be called belief and reality: The reality always was that the brahmanical people held the skill of writing, whereas others, especially the dalits, did not. This is not very different from how (following the essay "Structural Study of Myth" by Lévi-Strauss), ancient Greek culture attempted to resolve the contradiction between belief and reality through the Oedipus myth. This allows us to see for ourselves that it is actually the dalits who were, and still are, far more likely to have an "oral" tradition, which is precisely what is at stake: By claiming orality for their own tradition, the brahmanical people deny any tradition whatsoever to the dalits, since the only tradition the dalits could have had was an oral tradition. It should be noted that we are not making the often-made distinction between "primary orality" and "secondary orality" (e.g., Walter Ong makes this distinction in his *Orality and Literacy*).[9] In raising the stakes thus, we have the benefit of seeing how the brahmanical tradition *appropriated* "orality" for itself, thus denying the dalits any tradition whatsoever. We believe that the rise of dalit literature must be interpreted through this as well. An early dalit thinker, Jotirao Phule,[10] seems to have read aloud his books in several gatherings. Beginning from Phule, then, one can see an increasing dalit use of the written traditions, culminating in the 1970s into "dalit literature."

A lot of work on the history of caste has concentrated on the genesis of caste, whether in terms of a prior unity that diversifies into caste or a diversity slowly accumulating around the kernel of the brahmanical "Aryan" order. There are, naturally, other historical explanations of the origin/s of caste. We suggest that what is needed is not a study of the origins of caste, but a history of the practice of caste, of caste in the everyday life of people. Without doubt this is a daunting historical task, but we are of the opinion that it is one of the most important historical tasks, and possibly the most rewarding. Lacking the training and/or skill to undertake this task, we have approached the issue in a theoretical manner. Yet it is possible to imagine the initial contours of that task.

Since historical information on everyday life is lacking, let us imagine historically. Let us imagine everyday life in, say, the eighteenth century. It is clear from most of the available accounts that, in this period, as in earlier ones, dress itself was a matter and marker of caste. A lower-caste woman wrapped the sari around herself in a style that was necessarily different from the way a brahman woman wrapped the nine yards of fabric around herself. The same is true of men wearing dhotis and some form of headgear. It is also well known that the "untouchables" had to wear explicit marks of untouchability on their bodies—visible signs such as brooms tied to their backs and audible signs such as small bells. Thus, in the period, the question of needing to know, or knowing, a person's caste could not arise, because caste was clearly marked on the person. Caste became known through visual, and nonlinguistic auditory markers. Presumably, a child's body had to learn to acquire these markers, and a child had to learn to identify certain visual phenomena as markers of caste. We could safely say that this was learned first and the knowledge of the essential attributes of this or that caste was acquired later. Caste manifested itself, became phenomenal, through this mode of visuality; it first occupied the phenomenal perceptual space.

This mainly visual organization of perceptual signs of caste needs some consideration. Inasmuch as caste was always predicated on birth and was, on the whole, a matter of being from which all the doing and the saying and the feeling was supposed to derive (X speaks like that "because" he or she belongs to Y caste, and he or she belongs to that caste "because" he or she was born in it); the caste should have been, in principle, self-evident, written, as it were, on and into the very being of a person. However, this ontology of the sociality of caste, to put it awkwardly, needs to disseminate itself through a whole system of phenomenal visual signs, and through speech patterns as well. This need, either for reasons of economy of identification (how would one perceive the predication on birth without getting into "expensive" conversation or genealogical investigation?) or for phenomenal, material bodily expression, allows us to posit a fundamental difference between the predication on birth and its practical-phenomenal expression. This also generates the possibility of dissembling, deception, and other forms of the signification going differently from what the system expects. The naturalness of caste stands de-natured in the very markers of caste. It is therefore no wonder that it was these markers which were the initial ground of the struggle against and for the maintenance of caste: The "mahar" who refused to say "zohar maay-baap" to an "upper-caste" was said to be stepping across caste boundaries and was therefore "rude" and therefore needed to be "taught a lesson" and so on, as also the "lower caste" who refused to wear the broom and spit-pot.

The various historical changes in the style of dressing need to be examined from this point of view as well: The gradual disappearance of caste-marking items of dress or makeup resulted in what many people of those times must have perceived as an "anonymization," making social relations a little strenuous. (The relations between anonymization and modernity, and anonymization and individualization should not be forgotten here.) Now language had to be used, questions asked discreetly or otherwise, before a person's caste could be identified, and relations adjusted proportionately. Within language use too changes were taking place. Phatic expressions, which were quite often indicative of caste (a mahar always said *zohar* to an upper-caste person as a phatic introduction to his speech), now did not always indicate caste. If someone merely said "*namaste*,"[11] you did not get to know this someone's, this person's, caste. With the gradual disappearance of visual markers of caste, speech itself begins to be burdened with the task of identifying caste. Now one listens more carefully to speech patterns, accents, register, vocabulary, and so on because one wishes to identify the caste of the speaker.

It always was possible to identify caste with these; however, it was not necessary earlier, because the marks of caste were always visibly present, and usually before auditory linguistic marks—one identified caste without having to speak. One becomes sensitive to language in a new way now. The earlier regime of ways of knowing caste is undergoing a change. It is therefore not surprising that there is a specific relationship between dalit literature and caste as such, which has been misunderstood until now. Dalit literature was seen as the literature written by born dalits. However, the issue is not that of birth, but of speech patterns and their use for identification of caste in society. The strength and vehemence of dalit literature stems not merely from the pride in being dalit, but more so from the assertion of the validity of these speech patterns, bordering on the assertion of the right to express and/or record lived experience. We mention this issue briefly now and will take it up later. We believe that the brahmanical insistence on purity and clarity of enunciation must have acquired the function of being a caste mark in this period. A mark that always existed as a possibility within the semiological framework is now actually deployed and with this specific function of identifying caste. With that short digression, let us return to the issue of texts, and their place in society, and the analysis of that place.

The nineteenth century is the period in which there is, through the scholarly and/or officious and possibly tangentially imperial intervention of the Orientalists, a rediscovery of the great Sanskrit texts, which are quickly translated, edited, and reissued. This coincidence of the gradual disappearance of obvious markers of caste and the rediscovery of Sanskrit texts makes it possible to decry the disappearance on the basis of Sanskrit texts. Caste itself is, in a certain sense, rediscovered in these. What we mean is that the rediscovery of ancient Sanskrit texts made the brahmanical order realize that certain kinds of caste boundaries had disappeared and made it possible for the brahmans to complain about this disappearance or mourn the loss of ancient "standards." This could be blamed on two processes: the imperial, Orientalist, British intervention (conscious or unconscious, explicit or implicit) into religious practices or the processes of modernization.

We must note, along with the above, that the rediscovery of ancient Sanskrit texts on caste made them available to dalit thinkers as well, few as they were. This is also the time, roughly, when education made it possible for dalits to access the newly rediscovered Sanskrit texts, along with Western texts, and challenge the views presented in them. The best and possibly the earliest systematic challenge is to be found in the writings of Jotirao Phule, especially his *The Cultivator's Whip-Cord*, chapters 3, 4, and 5. Once the dalits themselves could read the ancient texts, the argument about caste could be held by two parties who could source texts that were available to both—in short, a real textual and interpretive argument between dalits and nondalits could develop only after this period. The equitable availability of the religio-metaphysical texts, as well as the epics and other texts, generated an odd, an almost aporetic, situation. On the one hand, inasmuch as the dalits wished to argue against these texts, the authority of these texts had to be granted first in order to be put under question later; thus the texts formed the framework of argument. On the other hand, dalits had to question those very texts that formed the very framework of the argument, thus making argument as such impossible. Also, since contemporary caste was said to be derived from the authority of these texts, dalits had to search for their own historical archives, which often enough were unwritten, unseen, and unacknowledged. The historical accounts in Phule's *The Cultivator's Whip-Cord* (chapters 3 and 4) again serve as the attempt to retell the history of dalits. We discuss some of this in greater detail in the next section.

The semiology of caste underwent a change, but we do not mean to suggest that the caste structure itself did not change. This is the resistant core of the discourse of caste, and of the phenomenon of caste, of the inability of sociologists and analysts to think of the changes that took place in the structure. We suggest that the changes in the semiology must be read carefully, even symptomatically, in order to look at the changes in the structure. The change in the semiology suggests clearly that the material and substance on which caste was predicated was changing: From bodies it was changing mainly to speech and, equally important, food habits. We will not discuss food habits here because they serve, till today, as convenient and clear markers of caste distinctions, and need separate attention. But it is clear that a regime of caste distinctions that is founded on bodily caste markers and one that is founded on speech and food habits must be different. Without doubt, even at this time, birth is still the most certain marker of caste. We have discussed this founding of caste on birth in some detail earlier, and now we are paying attention to higher-level distinctions. We notice here a curious interaction between the determination of caste by birth, and the determination by other markers—the former is seen as invariable, whereas the other is seen as variable. The anxious persistence of a metaphysic is seen here operating in society—that which you are born with is permanently you, whereas that which you come to invest in, or divest yourself of, in this case, the visual and speech pattern as caste markers, comes to be seen as impermanent and capable of misuse or misrepresentation. It is in this interaction that the genetic fixity of caste is hypostatized into a transcendental principle and explanation of caste as a social, historical, variable practice. The genetic fixity of caste is deep-rooted and can be found in anti-caste writing and political activism. It is not without irony that large parts of the history of the anti-caste struggle show a concentration on the more or less semiotic markers of caste, and not on the genetic foundation.

In the process of modernization, it was the visual markers that disappeared first. We should be clear not to confuse the look of the people and visual markers. The racial interpretation of caste (where the more "dark-skinned" and "anthropoid" are the features of the face and the body, the lower in caste one is supposed be) is distinct from visual markers on the face and the style of dress, over and above which there could be other markers like the spit pot or the broom tied on the body. At a subsequent stage of modernization, these detachable visual markers will turn into a nondetachable, "racial theory" of caste, and Western anthropology will assist in inscribing caste onto the very skin color and anatomy of people. It is now possible to say that the very sight of a dalit person will tell us that she or he is a dalit. These anthropological discourses will render caste into a nondeceptive, integral, undetachable, natural phenomenon: Anatomy itself becomes a perceptual marker (unlike the earlier times, where deception was always possible). Supporters of caste must now depend on these to speak of and practice caste, since the obvious markers are gradually disappearing. Caste must now find support somewhere else, no longer finding easy support in actual social practices of marking caste.

With this, we bring to a close our discussion of classical texts, concluding, as indicated earlier, that these are not useful for understanding or describing the practice of caste as such, whatever force of legitimation they might have had or still have. It is not only the metaphysics within which these texts embroil us but also the specific deployment of classical ideas that needs to be questioned and challenged. This could happen only after written texts cease to be sacred and/or secret objects and through the agency of the printing press become commodities that are addressed to anybody who can read. Once such texts were printed, the possibility of challenging and questioning their contents becomes available.

Possibly the earliest systematic challenge is to be found in the writings of Jotirao Phule. Much has been written on Phule, and later on by and on Ambedkar, and in the next section, we pay attention to modern texts and their place in society. It can be argued that we have summarily dismissed the ancient Sanskrit tradition, and that would be a valid argument to a certain extent. However, as indicated in the beginning of this chapter, the attempt is to make a conceptual argument, and not an argument about individual texts, or even a comprehensive one. If the logic of our argument is convincing, then pointing at some ancient Sanskrit texts that are against caste does not refute the argument.

THE PLACE OF TEXTS IN SOCIETY AND ITS ANALYSIS: MODERN TEXTS

In the preceding section, we discussed how classical (ancient Sanskrit) texts function in society and how the textuality of visual markers of caste operated in earlier periods. In this section, we discuss modern texts. As has been our practice, we concentrate on Maharashtra. We focus here on dalit texts, since it is to be easily observed that more than anything else, it is the entry of dalits in the field of writing that marks modernity in Maharashtra. It is possible to locate three moments in the history of this writing.

Without doubt, Jotirao Phule's writings inaugurate the modern anti-caste struggle, and we must pay attention to the features of this modernity and the features of the argumentation. It is not unremarkable that Phule's attack on caste must necessarily begin with a rewriting of the history of caste (especially the narrative of the formation of caste) and the history of the "shudra-ati-shudra" (the lowest of the low shudras). This specifically is a rewriting since it seeks to de-legitimize the existing brahmanical accounts of the formation, practice, and justification of the practice of caste, of untouchability, and the exploitative intent of that justification. *The Cultivator's Whip-Cord* is, from this point of view, a key text. The organization and partitioning of this text needs some comment. The first chapter describes the nearly diurnal practice of eliciting from the lower castes, through all manner of deception and cunning and ritual pieties, rather small amounts of money. The third chapter describes the origins of caste and a history; the fourth describes the contemporary everyday circumstance of life of the impoverished farmer, with the fifth making suggestions for changing the situation. Phule's writings are powerfully polemical and colloquial. After the so-called "saint-poets" of the bhakti movement, it is only in Phule's writings that we find a completely unselfconscious use of nonbrahmanical Marathi. The major difference between the nonmodern writings of the saint-poets and Phule is the near-contempt that Phule shows for any brahmanical metaphysical and ritual ideas. The saint-poets were still trapped in notions of god-hood, which, though radical enough, were not totally different from some versions of god-hood available in the Sanskritic metaphysical traditions (making it possible for them to claim that the impulse of bhakti is as old as Shankaracharya at least, mentioning his four verses in the "bhakti" composition "bhaja govindam," if not as old as the *Upanishads*). It is also noticeable that Phule does not derive his ethics from Hinduism, whereas such derivation is the mark of the writings of most saint-poets.[12]

We have earlier indicated the issue of language and speech and the issue of legitimation. Phule's diction shows that for him the question of any anxiety about his own language does not even arise. His rhetoric is, quite often, that of a street-polemicist, but the arguments that he offers are, in the final analysis, always independently, and in a non-Hinduist manner, ethical. Phule's political actions are well known, but his writings are not equally known; a lot of them are untranslated, or badly translated. Therefore, we concentrate on his writings and not his political action, retrograde though doing this might sound. It is perhaps not unremarkable that the first outburst of the anti-caste sentiment and politics expresses itself as difference in style. A practice must articulate itself in language, often before it sees itself as a political practice. What is remarkable about Phule's writings is how his style explodes into Marathi. It is well known that the kind of Marathi that Phule uses was unacceptable in varying degrees to the more educated people. The prose style does not have very many precedents. Even if Phule wanted to draw on a tradition, there was no historical tradition available— he could only draw on his own language (with its own history).

Phule's argument, especially in *The Cultivator's Whip-Cord*, is easily summarized (because it is so lucid). By making the shudra-ati-shudras believe in Hindu religious mumbo-jumbo, the brahmans convince them of the necessity of certain rituals and make money or food out of these rituals. (Even before a shudra farmer is born, there already is ritual; after he is dead, there still is ritual. Phule gives a rather detailed enumeration that covers almost every day of the year, of the various rituals that the shudra farmer is supposed to perform.) The brahmans also possess knowledge and education, and thus, in the modern regime of earning money through employment with the British administrative and civil authorities, the brahmans monopolize the alternative sources of income that might have been opened up to the shudra-ati-shudras. Moreover, the brahmans also ensure that these never get a decent education. This is how the brahman hegemony is produced, and, more important, this is how it is maintained. The behavior of the British officers is complicit: They either are busy with their own colonial comforts and luxuries and, in a few cases, debaucheries, or they do not care enough. Those who claim to care are entirely dependent on brahman informers and interpreters and therefore do not ever get a clear picture of the condition of the shudra-ati-shudras. The same applies to all the brahman do-gooders and reformists, since they never attempt to reform caste itself. It should be noted that Phule also exhorts the shudra-ati-shudras to educate themselves, to get out of the situation they find themselves in.

In a certain sense, Phule's argument seems to be that the brahmans always controlled or guided state resources. In the pre-British times, they were always in key administrative positions in the various kingdoms, princely states, or jahagirs, and in the British times, they consolidated their social position by quickly securing employment with the British. These last, in turn, were either naïve or careless, in spite of their general belief in ideas of equality. Relying on brahmans as their "native informants" as they did, a clear picture of the condition of the shudras was never available to them. In limpid, polemical, but moving prose, Phule suggests that some of the British officers' wives could do well to visit a shudra marriage ceremony and see the amount and kind of filth that it generates.

By controlling and guiding Hinduism, the brahmans circulated an incorrect version of history, and therefore an "alternative" (meaning "correct") history of caste became a necessity. This history must begin at the beginning. Allow us to quote a remarkable passage, albeit in translation:[13]

> In all this incomprehensible, unthinkable, ether-like space, countless solar systems come into existence and pass away, through the joining and separating of elements. In the same way, when every planet is following its own sun, on this planet earth, through the uniting of the same father and mother, different children are born, one intelligent, the other stupid. So on this evidence we cannot infer that intelligence and stupidity are hereditary . . .
>
> Now if we say that according to Darwin's notion, following the movement of planets in the whole solar system the animal monkeys transformed into the different species of the new human beings, then this harms the notion that humans were created from the limbs of Brahma. Therefore, while negotiating with all these notions if we now say according to Buddhist or Jain notions that human beings were cre-

ated from pairs, or according to Darwin we say that men and women were created from monkeys, or according to the Christian notion that God created men and women from clay, or according to the brahman notion that the four castes of human beings were created from the brahman's body, and move ahead; then we will have to say that when men and women were created, they must have had to live under trees, or in tree-hollows, or in caves, and must have had to eat roots and fruit from the surrounding forest; and when in the after-noon they might have rested under the shade of tree, away from the bright sun, they must have seen really high mountains with broken ridges, and cliffs, and rows of peaks as if wearing white caps of mist, and below them in the valleys, huge trees, like banyan, peepal, and trees of jack-fruit, mango and coconut, fig and pistachio and almond, all humbly bowing with the weight of fruit, covered with nets of creepers and grape-vines, and in many places laden with bananas, and multifarious flowers like lotuses, and all around them the ground covered with leaves and flowers, and an immense vari-colored carpet formed, and on it, trees laden with leaves and flowers, as if newly planted. Also, cucumber, and melons of all kinds recumbent in all the streams and rivulets and river-beds, and everywhere the clear water flowing with wondrous and interminable melody. Nearby, on the faces of lakes, swarms of bees humming over the colorful lotuses, and in places, on the shores, herons, as if meditating on one leg, waiting for insects to come within reach. In the forest close by, herds of poor deer and sheep struggling for breath as they run hither and thither to save themselves from beasts like foxes and tigers, and in the trees, a variety of birds singing melodiously enough to put Tansen to shame, and when they are so engrossed, high above in the sky, birds of prey like hawks and falcons circling and waiting to swoop down and take their lives. And then, the soft and cool breeze from the west has brought the scent of a myriad flowers and rendered everything redolent. Seeing all this, how joyful must have been the original ancestors of humans, those who now call themselves Buddhists, Christians, Muslims, Mahars, and Brahmans! So be it, but must not they have had to live naked, with long nails and longer hair, because they did not possess the knowledge to make tools, or clothes? Not knowing how to make pots from clay, or metal, they must have had to drink water from the river, bending low like animals, or with cupped hands! How would they taste chapatis, if they did not know how to make pans, or grinders? Must not they have had to walk barefoot, not knowing how to skin sheep or cattle? How would those who would have difficulty in counting up to hundred, under the influence of Somarasa, know how to roast and eat cattle? In sum, they must have been so ignorant, that if some charlatan or cunning person had put before them the book of the vedas, cut on parchment, they would have taken it into their hands, and seeing that it did not have any smell or taste, it is unthinkable what they would have done to it. Because they were fruit eaters themselves, would they, as these nocturnal beasts the makers of the Vedas say, have been able to steal cows and eaten them, under influence of Somarasa, or for the shraddha, and would they have needed to? Because they would have been so pure, would they have liked to call these selfish writers their inheritors? Would these writers have had the gumption to stand before them and say, "You're Buddhist," "You are Christian," "You're Mussalman," and "You're low because a Mahar," and "We are superior because we are Brahmans"?

Phule goes on to recount how the brahmans defeated the dasyus, and how eventually the powerful shudras were decimated, and how those who survived were emaciated, and later gave their consent to brahman hegemony. In this history, Phule portrays the mughals and the Christians as agencies of some real and some possible liberation.

Given the argument, it is inevitable that Phule should feel the need to posit a version of religion as well. We shall not go into details of how his version of religion and religious reform is different from other attempts at reforming religion, for example, the Brahmo Samaj, and the various other "samaj"s that were formed in the period. The difference naturally turns on the issue of caste. Other reforms in Hinduism were either inattentive to the issue, or, according to Phule, merely made pious statements that were never supported by practice.

What stands out, demanding our attention, is the point that brahmans have appropriated state resources. This is a view that will repeat itself in subsequent anti-caste writings and politics; as is well known, even today, the question of reservation policies relates more or less directly to employment in, or access to, the various institutions in the several "state-apparatuses," to recall Louis Althusser's famous use of the collocation of state and apparatus.

Perhaps it is in Phule's writings that we find a very early critique of inheritance. In a certain Marxian sense, we could say that the form of social relations is lagging behind, while the content has run away with itself: Several artisanal occupations have become useless (unproductive) even for mere sustenance, and people are now taking up other occupations which they never inherited. But in social practice, it is still inheritance that defines social position and place. We have already discussed societies of inheritance and societies of acquisition. In the transition from the former to the latter, the specific experience of the shudra-ati-shudra was that of dispossession. It was upon this dispossession that they built their acquisitiveness, whereas for the brahmans, it was an experience of addition: Acquisition added to inheritance. Several brahmans transformed themselves into industrial entrepreneurs, or, to speak like Phule, wormed themselves into the state bureaucracy. This is not to deny that quite a few individual brahmans were dispossessed; however, compared to the recursive dispossession of the shudras, their numbers were smaller.

With Phule, and his followers, a new discursive situation obtains. "Caste" has become visible in the life of society, no longer as the principle of social organization but as a problem in society, more specifically, a problem in Hindu society (though not restricted to it). Reciprocally, the notion of equality becomes equally visible as desirable. In this process, it should be noted, that unlike the notion of equality before god, this notion of equality always assumes a fully formed individual agent, in short, a subject who understands himself or herself as a subject. However, in this period of transition, especially for the shudra-ati-shudra people, the access to such a subjecthood was not really available in terms of real buying power, or real rights, and those subjects who understood themselves as subjects could not really practice their subjecthood.

This is the period, time, duration, when a resistant group of people desire that which they resist. There is, in principle, no reason why the resistant, perhaps protestant, group might, or might not, want to do what the dominant group does. However, we witness here the transition—ever so fragmented—from inheritance to acquisition (modernity we define as the business of acquisition).

The second moment is of the other modernizing dalit author, B. R. Ambedkar. He has written much; his collected works running into several volumes. While this is not the place for analyzing his writings, it is necessary to state that Ambedkar inaugurated the second moment of modernity, and was, in the final analysis, far more influential than Phule. His status as an activist, and as a thinker, sociologist, articulator of law, leading contributor to the Constitution of India, has often obscured his own analysis of the caste question.

Like Phule, Ambedkar too expects the state to intervene in the question. We have discussed elsewhere some of this; therefore, we give here only a sketch. The much-discussed "reservation policy" was designed precisely for such an intervention by the state. It follows, therefore, that Ambedkar did not think that the caste question could be solved in a civil manner (though he does not explicitly make the distinction between civil society and state). It also follows that the notion of the relationship between the state and the people was envisaged differently from other political thinkers, Western, or Indian. It may not be an exaggeration to suggest that the concept of the state itself was different in his writing. The other attempt (over and above the various political actions and campaigns like temple entry) to step outside of Hindu religion (and metaphysics) was to convert to Buddhism. Needless to say, like other claims to stepping outside, after the conversion the outside very quickly repeated and reproduced the inside that had apparently been left behind, especially in the decades that followed Ambedkar's death: Those who had converted to Buddhism retained several "Hindu" practices.

Moreover, we must wonder if the only way to step out of Hinduism is to step into some other religion, in this case, no doubt a very considered step into Buddhism. We must wonder why the option of atheism was not thought to be available.[14] It seems that both these moments are moments of defeat and failure, in a certain sense: defeat at the hands of the Hindu orthodoxy and the Hindu people, and failure to think through the question of the inside and the outside. It is also to be noted that in *Annihilation of Caste* Ambedkar has recourse to precisely the type of "anthropology" that had successfully racialized caste. Although it is true that a thinker can only work with forms of knowledge that are available, it does seem a little odd that Ambedkar fails to see through the racializing discourse.

A large amount of Ambedkar's writing is in English prose. His relationship with Western notions is as complex as Gandhi's, if not more. We must pay attention to issues of language—in the sense of *which* language is being used. In Phule's case, it was his *dialect* of Marathi; in Ambedkar's case, it is a different language altogether. Ambedkar has written in Marathi as well, but he does not use a "dialect." Ambedkar is philosophical in his own way, and we quote from the blurb of the seventh volume of his works published by the Government of Maharashtra:[15]

> To idealize the real, which more often than not is full of inequities, is a very selfish thing to do. Only when a person finds a personal advantage in things as they are that he tries to idealize the real. To proceed to make such ideal real is nothing short of criminal. It means perpetuating inequity on the ground that whatever is once settled is settled for all times. Such a view is opposed to all morality. No society with a social conscience has ever accepted it. On the contrary, whatever progress in improving the terms of associated life between individuals and classes has been made in the course of history, is due entirely to the recognition of the ethical doctrine that what is wrongly settled is never settled and must be resettled.

Ambedkar's failure was a failure of the state of anthropological and ethnographical knowledge of his times (which would essentially mean the formative times of his life, irrespective of his "age"). The failure is obvious in his recourse to anthropometry and ancient classical texts, and their internal contradictions, or the clues they provide him for a rewriting of a history of the shudra.

In his writings, however, there are anachronistic Foucaultian moments. We shall soon quote the relevant passage, but we beg indulgence to digress into our memory as teachers. Think of a classroom. A child has written something "dirty" on the blackboard. The teacher comes in, sees it, and wants to know "who" wrote it. Not one of the students owns up. The teacher gets furious. Time passes. The teacher gets more furious. Students are silent, still. Finally, the teacher says, "Unless I am told who wrote this here, the whole class will be punished. To begin with, it will recite Milton."

The class is silent for a while, but soon, in less than a minute there is divergence, dispute, blame. This is what Ambedkar writes, 187 pages into his argument:[16]

> In all ancient societies the unit of guilt was the tribe or community and not the individual, with the result that the guilt of the individual was the guilt of the community and the guilt of the community was the guilt of every individual belonging to it. If this fact is borne in mind, then it would

be quite natural to say that the Brahmins did not confine their hatred to the offending [Shudra] kings, but extended it to the whole of the Shudra community and applied the ban against Upanayana to all Shudras.

The third moment is that of dalit literature of the Dalit Panther period. Again, the question is mainly of language, this time, a return to the question of *dialect*. The explosion of this literature was powered by a number of dialects. The authors, men and women, do not seem to need any outside legitimation for the use of their own dialects. The seed that Phule had planted in Marathi has, by this time, become a full and many-leaved tree. It is clear that this linguistic confidence also bespeaks other kinds of confidence, especially a confidence in their own ethics.

Given these texts, we must consider what their position was in the past and what their position is today. It would not be an exaggeration to state that the majority of shudra-ati-shudras are illiterate and do not have access to these writings and the ideas they contain, except through some interpreters. Nevertheless, a textual community does come into existence: "Folk songs" are composed, sung. Phule and, later, Ambedkar become deified in some circles, literate ones, deified enough to prevent a critical understanding and appreciation of their writings.

We indicated in the beginning of this section that dalit writing is the best marker of modernity in Maharashtra. This is not merely a matter of "voice" and the various actions that can be performed on it (giving, taking away, silencing, etc.). This is also about formations of identities and formations of politics. The two should not yet be collocated into identity politics, despite the contemporary debate around this collocation, for identity must involve itself into some metaphysical notion of identity whereas politics must seek its energies in ethics. This is not to say that identity politics do not exist but to say that they may very well mask actions and aspirations. It may be premature to say that the notion that the politics of identity and the politics of recognition (both often seem to go hand in hand) explain the political practices of people. When we seek a definition of recognition of our own identities, there is a double movement: identity for ourselves, and our self-recognition, and identity for others, and the need to be recognized by others as this or that identity. What is modern about dalit writing, then, is not merely the well-known fact that this is the first time that dalits are writing, but that the fact that dalits are writing is the mark of the inauguration of the transition to modernity.

SOCIETY, TEXTS, TRUTH

Society is circumscribed by caste, in the sense that members of a caste consider only other members to be legitimate members of their own caste-society; others are seen as "outsiders." Relations with outsiders are seen as mainly instrumental relations, insignificant for social life, since they are not kinships (there is no commensality or connubiality). It might seem that narratives such as the *Ramayana* and the *Mahabharata* actually connect these caste-societies, since these narratives are part of the textual fabric of the totality of caste-societies, though they change and are disseminated in a variety of ways. However, the textual operation of these narratives is more complicated, since these often serve as moral texts from which one is supposed to derive one's own morality. This ideological operation that renders material relationships into ideational moral relationships needs to be taken into account before it can be said that the various appropriations of these narratives are fundamental to sociality. Nevertheless, it must be stated that these narratives—and others like them—constituted a major source of knowledge on how to behave in given situations and perhaps could even function as precepts. It is not easy to estimate the social power and behavioral efficacy of these narratives, but it is clear that they circulated widely and informed the moral universe of most parts of caste-societies. These moral precepts could always be given what we might call a custom override, and our hunch is that custom prevailed more often than idealized moral precepts.

Before the invention of the printing press, the texts were performed in spoken language, and although it is possible that they were commodities (in the sense that they were exchanged, expressing a certain exchange-value), their circulation largely depended on the circulation of people who had memorized them. Once the texts are printed, the memorization gets housed in the external memory called a book. We are, for the sake of brevity, working with commonly accepted usages of the words *internal* and *external*. It is necessary to admit that the distinction between internal and external memory is always made, but it is difficult to identify the way it is made. If we take this into account, then the difference between writing and printed writing also becomes clear. Although both are subject to all the vagaries that writing is subject to, they are subject to these vagaries and philosophical issues in different ways. Elizabeth Eisenstein's work on the printing press as an agent of social change[17] is one clear demonstration of the difference, although it is undeniable that both writing and printing function as external memory. It might be necessary to wonder if printing is a copy of writing, or whether they are as different as speaking and writing.

The other feature of the place of texts in ancient and modern societies is their authority by the fact of being texts. It is to be remembered that in ancient societies there is an interaction between authority and texts—those texts that have authority will be preferentially "textualized," that is, rendered as written texts, or stabilized through mnemonic techniques; and in turn, such written or stabilized texts will gain and retain authority. In modern societies, especially print societies, many more texts are printed and stabilized—there is a proliferation of texts through printing, translating, and so on. Such proliferation necessarily takes away some of the authority of "authoritative" texts. Minimally, their authority is de-sacralized, rendered available for human contemplation and manipulation. Also, since the method of production makes mass production of books possible, their circulation too undergoes a proliferation, as does their relationship with readers—this has been already discussed by Walter Benjamin in "The Work of Art in the Age of Technological Reproducibility," and several subsequent discussions have taken place around the themes and issues first discussed by Benjamin. We can only take these discussions as read, just as we can only take as read discussions of "text," "textuality," "inter-textuality" in the second half of the twentieth century—we have Roland Barthes's and Jacques Derrida's writings in mind. In ancient societies, the authority of texts marks them as truth-bearers, or truth-carriers, whereas in modern societies truth proliferates and multiplies, it loses its special unitary character. Texts in modern societies can, on the one hand, add to the truth, supplement it, revealing its original inadequacy by supplementing it; on the other hand, they can be made into purveyors of some absolute empirical truth that is taken as scientific. It must be remembered that when we say that truth has a unitary character, we are speaking of caste-societies, in which particular texts have authority. This is evident for example in the various recensions, versions of "authoritative" texts like the *vedas*, the various *dharmashastras*, and their position in the various subcastes within the brahman caste.

Yet another feature of texts in ancient caste-societies is that they could be objects of the barter form of exchange. In modern caste-societies, they become objects of the commodity form of exchange, and, as already pointed out, proliferate through mass production. Although it is true that Walter Benjamin's essay and subsequent discussions have made it possible for us to understand the mechanisms and social implications of mass production, perhaps it is worthwhile to ponder a little over the identicality of mass-produced objects and their proliferation. The identicality could be understood in terms of imitation and repetition, the proliferation through statistical processes of the kind that Prigogine and Stengers discuss in their *Order out of Chaos*,[18] both supplemented by a consideration of Gabriel Tarde's treatment of "imitation" in his *Laws of Imitation*.[19] Each printed book is a repetition and imitation of other copies of the "same" printed book: Mass production mechanizes and generalizes and proliferates repetition and imitation, just as it releases new energies by increasing access to otherwise restricted knowledges and histories. A new possibility arises through these processes: the possibility of repeating knowledge and truth articulated in texts. New commonalities form across caste-societies, releasing the possibility of conceiving of society as such, that is, society beyond caste-societies. It is this nascent, yet-to-come society beyond caste-societies that becomes the driving force in the anti-caste struggle. The new truth of caste-societies as unjust societies is articulated. We can safely assume that along with new truths, new fictions too arise.

THE CONSTITUTION OF DALIT LITERATURE: TRUTH AND LITERATURE

As more dalit people enter the domain of the literate and the educated, the domain gets modified, even if not transformed. Inasmuch as written and printed material is addressed to anonymous readers, it becomes possible for dalit authors to address nondalit readers. The role of print capitalism should not be ignored here. It is precisely printed material as commodity that opens up this domain for dalit authors. In a society that already is literate, the power of this written and printed commodity increases in manifold ways.

If earlier certain kinds of writing were the prerogative of brahmans, now that kind of writing can be practiced by dalits as well. It is an interesting aspect of the history of Marathi literature that brahmanical literature invoked the ancient Sanskrit past as its historical context (as literary resources), whereas dalit writing had to negate precisely that historical context. The nineteenth- and early twentieth-century history of dalit writing did not provide much by way of a context for fictionalizing or poeticizing, since most of dalit writing then was either political or social. The only literary context was stories that were inherited in what necessarily had to be a primary oral tradition, and these were not particularly conducive to written or printed literary narratives or poetry.

Negotiations between this new literariness and primordial storytelling can be seen in the styles of such early writers as Shankararao Kharat, Baburao Bagul, and P. E. Sonkamble. We focus on Baburao Bagul, in whose writings one can see some interesting enabling contradictions. Since we have discussed this in some detail elsewhere, we shall quickly summarize it here.[20] Two basic characteristics of Bagul's writing stand out: (a) a "literary" style that is dependent on such ancient Sanskrit norms as were learned in his literary career (not necessarily through formal education), and (b) characterology and ethical complexity and narration of suffering. Inasmuch as the diction is replete with Sanskrit and Sanskrit-derived Marathi words, and often alliterative, and the narration is of the suffering of characters who do not have access to such diction, the stories and novella are pulled in two opposing directions.

Since alliteration is a frequent figure, perhaps a literary critical look at the figure is needed. Alliteration functions on the principle of morphological revaluation of the values at the phonetic level: The initial allophones are "the same," but the rest of the word is different—the figure functions on the vertical axis of integration. We borrow this notion of integrational axis from structural linguistics—Kartsevsky, Jakobson, Trubetzkoy, and above all, Roland Barthes's "An Introduction to the Structural Analysis of Narrative." Jindřich Toman's *The Magic of a Common Language* provides a useful account of how in the discussions among its members, the Prague Linguistic Circle moved away from the relatively simpler notion of the differential relationship that was proposed by Saussure in his *A Course in General Linguistics*, first in a talk given by Kartsevsky and then in the correspondence between Jakobson and Trubetzkoy.[21] They began to appreciate the fact that differences at one level—let us say the phonological level—are "integrated" at another level—the morphological level. Alliteration as a figure of sound functions precisely in the verticalization of identity and difference: "the murmuring of innumerable bees." It is in the third occurrence of the sound "mar" (innu*mer*able) that it is revalued at the morphological level. The difference is that of naming: What Saussure had called "paradigmatic," Jakobson "associative," and Barthes "integrational," we would like to call "reevaluative."

It would not be an exaggeration to state that in Bagul's writings the diction is fully revalued at the level of plot and character, and that this is a persistent feature of these writings. Bagul needs to be seen as an author who made the discord between style and content an act of constituting the literary-aesthetic. In this sense, Bagul could be said to be a modernist author—provided we do not stop analysis after merely identifying the diction as being replete with Sanskrit or Sanskrit-derived Marathi words. The characters, who usually are deprived of all that even a lower-middle-class value system takes for granted, often are placed in liminal places and situations. This too is a revaluation, since such characters had hardly entered the realm of literature in Marathi. We shall not enter here into a discussion of Marathi literary history, although we think it necessary to point out that the fundamental change that took place in Marathi literature was in the nineteenth century. All earlier literature had plots and characters derived from epics, *puranas*, Hindu/ Mughal mythology, and other preexisting repertoires, reservoirs, resources, stores, and magazines of narratives. The idea of telling an imagined story with previously unknown characters and events came up only in the nineteenth century.

From early novels like Ketkar's *Brahman-kanya* (1930s) on, and through periodicals (some of them run by B. R. Ambedkar), "caste" came to be written about within the field of the literary. A truth that had already found its articulation in nonliterary writing came to find a place in the literary as well. It is to be expected that such literature will have a special relationship with "truth," and that this will be different from other literatures that have their own relationships with truth. If we focus on this special relationship with truth, dalit literature begins to look a little different from what its mere popularity might make it look like, and from the nondalit, privileged people's consumption of it—to signify their own modernity and liberality—different as well from the vague politics that have come to be associated with it.

We will have occasion later to discuss the politics of caste-societies; we shall focus here on the literary.

Fiction written around the theme of caste articulates a truth in the literary domain, a truth that already had found its own discursivity in European sociology and anthropology, and earlier descriptions of Indian culture in general. Not just Al-Biruni,[22] or Abbe Dubois,[23] but also Ziegenbalg,[24] not just the early travelers and the orientalists, but also European, and later specifically British and Indian administrators. In this context, we must quickly discuss the sensitive issue of the colonial period and what the British did to Indians. To begin with, this dualism itself has to be put under question, for the story—the historical account—should be of what the British did to some Indians and what these Indians did to other Indians and what Indians did to the British who benefited from being here or not being here. It must never be forgotten that while there are glorified accounts of how the British finally defeated the "mutinous" armed Indians, the historical truth is that these British led some Indian soldiers who defeated other Indian soldiers.

The new writing brought a discursivity to the problems of occupations that were displaced and replaced, dislocated and relocated in the nineteenth century. People felt the need to talk of it, to write about what they had undergone—the easiest thing to do when being dislocated and relocated professionally, or, for that matter, while traveling. We must remember that the dissemination of skills through print had not yet proliferated. To enskill oneself in other professions required great labor, intellectual and manual, and an even greater will to live on. Europeanized education did play an important enabling part—but without the skill to learn something new, none of the members of caste-societies would have lived on to produce new truth. An indication of this skill is the well-known fact that many people who were introduced to European ideas did not, in fact, just "give in" to standardized liberal and middle-class ideologies of the Europeans. For example, Phule likes Tom Paine, rather than J. S. Mill, or Jeremy Bentham—authors much more likely to be easily available, especially since they already had become popular in Maharashtra; whereas the brahmanical Chiplunkar likes Samuel Johnson so much as to model his Marathi sentences on Johnson's English sentences. This choosing also allows our thought to move beyond the colonial and easy postcolonial notions that British and European ideologies were transmitted through education, and Indian intellectuals were merely recording machines, or even worse, blank matrices, or unwritten paper that received these marks. Moreover, in this line of argument, it is possible to follow Phule and point out that what is called British administration was actually, at the lower levels of authoritative execution at the level of performance of actions, filled with brahman intellectuals (who must have been the primary data collectors for the revenue and other administrative departments). The story is just not that the British changed us (the so-called "colonial construction" of gender, caste, education, ideology, caste, and so on): The story is that the British wrote out commands, and some Indians executed these commands, performing actions on the actions of some other Indians. Sitaram Pandey's account—in his *From Sepoy to Subedar*—of how he managed to avoid executing his rebel son in 1857 is instructive here.[25] One may think one is giving a command, one may think one is executing a command, but usually something else happens. This is an experience known to anybody who has worked within an institution. There is no reason to believe that it was otherwise in the institutions of the East India Company and the British government in India. However, our concerns are not to stop here.

We suggest that there was an engagement with the newly articulated truth of caste-injustice in the fiction that was written after the 1930s. If we take the writings of Baburao Bagul as exceptional, we begin to see that there is an intimate relation between truth and fiction: Here, fiction is in the service of rearticulating a truth that has been already discursively articulated as being operative in society. One would expect such fiction to be in the "realistic" mode, since there is also a certain "didactic" involved. This affinity between a "didactic" and "realism" is to be found in European fiction. However, as we have already observed, the style of Bagul's writing is not recognizably "realistic," and the plot of the novella *Sood* is not even plausible. It is also more than evident that the narration does not have behind it a conception of any "disinterested interest" or "nonpurposive purposiveness," it is not "aestheticized" in the post-Kantian European sense, neither in the American New Critical sense of an aesthetic object that is autonomous, autotelic, and so on. In fact, the clearly identifiable fracture between style and content serves, initially, to sensitize the reader to the gruesome lives that are narrativized, and later to initiate what could be called an ethical consideration, as the reader gets to read of the impossible performative situations in which characters find themselves. The reader reads through a narrative that highlights the ethical as an experience of an impossibility of being ethical, and yet one performs an action that will have consequences for oneself and for others. It is not merely accidental, we think, that many of Bagul's narratives end at what is often called the double bind. The ethical complexity is of such a nature that the author renders the reader impoverished in ethics—just as the characters are impoverished in ethical action. This ethical impoverishment is necessary, because it is only here, in this impoverished place, that a fusing of representation in the literary sense and representation in the political sense can be brought about. This fusing is a confusion as well, as subsequent history of the reception of dalit literature as explicitly "political" clearly shows—it ceases, almost, to be literary. The impulse of truth-telling that brought about this discursivity now fully realizes itself, culminating in the production and reception of "autobiographies" of the truth of individuals who are made to suffer because of their caste, as told by themselves. This confusion between the literary and the political, and the political and the ethical haunts the discourse of dalit literature. By now it is clear that authors, readers, critics, and political leaders have not been able to clarify themselves out of this confusion. The task of clarifying should have been taken up by the critics, but they have mostly found it more than ethically satisfying that dalit literature is a political and revolutionary phenomenon. Since the 1980s whenever political leaders have taken up these issues, they have found it more than satisfactory that they strive hard for implementation of the rules of representation and "reservation" as defined by the state from time to time. We shall not take up here what is often called the "numbers game" in elections, since we do not know anything about the nonpublic deals and secret negotiations between leaders and followers of various political parties and the several factions among dalit parties. It would not be entirely incorrect to suggest that it is not the idea of a democracy that is operative in these politics.[26]

It is because of the confusion mentioned above that the literary—in its most clichéd and rhetorical form—can be found in political exhortations for this or that political campaign, or electoral support. It is also to be noted that these political appeals and exhortations often invoke a sense of righting ethical wrongs, compensating for suffering and injustice.

The literary as one kind of imaginative work ceases to be as important as in other literatures, and truth-telling as the matter of narrating one's own experience becomes more important. As we have already indicated, this phenomenon has some consequences that are productive (the disruption of literature as an autonomous aesthetic production dominated by brahmanical conceptions) and some that are counterproductive, as seen in the confusion between the literary and the political, between the literary and truth-telling, between talking about one's experience and truth, between literary writing and political exhortation.

We have discussed the more specifically literary issues in "Destitute Literature"; therefore, we have here focused on what seems equally important, almost as if the ethical were the consequence of the literary, or, as is often claimed by dalit politics, as if the literary production were the consequence of the ethical superiority of victims of political and social domination, exploitation, suppression, oppression, and hegemony. It should be more than amply clear that the claim to ethical superiority (often mediated by literary representations) cannot compensate or account for political and social suffering. Dalit literature might have been revolutionary—but it should not be confused with revolution itself. For the revolutionary to be elaborated into a particular revolution (spontaneous or otherwise) requires large amounts of intellectual and political and social elaboration: Literature itself, however revolutionary, cannot perform this elaboration. Thus it becomes possible to suggest that just as nondalit readers of dalit literature read it to signify to themselves and others their own ethical correctness and generous sympathy, dalits themselves used the literary to signify their own ethical superiority through narrations of suffering. The powerful affects represented in this literature permitted authors and readers to confuse the literary and the political, the revolutionary and revolution. It now becomes possible to say that dalit literature is constituted through these various processes, of which all of us are a part, as much as Phule and Ambedkar used to be a few years ago.

6

(UN)TOUCHABILITY OF
THINGS AND PEOPLE

USING "UNTOUCHABILITY" TO DEFINE
THE PRACTICE OF CASTE

We have already seen that the traditional "scholarly" explanation of the origin and verifiability of caste uses the notion of "birth" (only a person born of a brahman couple is a brahman; those born of other conjugations—especially "intercaste" conjugations—are "lower," etc.), and we have also observed that this definition is operative in society today, on both sides of the anti-caste struggle, and has been for a long time. There are attendant identity politics that reinforce this definition, further consolidated by a metaphysics of experience and suffering, preserved and nurtured by an ever-increasing discourse and performance that reinforces these identity politics.

It is interesting to note that Jotirao Phule's writings do not seem to show him interested in this issue: He explains the origin of caste in terms of a historical struggle, as does Ambedkar to a certain extent. However, by the time we come to Ambedkar, things have changed a little. As is known, initially he attempted to define caste through socioanthropological terms. His 1916 presentation at the age of about twenty-five, "Castes in India: Their Mechanism, Genesis, and Development," explains the "genesis" of caste in terms of endogamy, and the differential treatment of "surplus" men and women, who are rendered "surplus" *not* through proliferation or production, in his account, but through depletion, death. Many later definitions also explain "caste" through various deployment of the concept of "kinship systems," which are various permutations and combinations of endogamy and exogamy. This is the case, to a large extent, with the early work by G. S. Ghurye, and later by Iravati Karve. "Birth" was not the focus of their writings, though what they historicized and sociologized and anthropologized was precisely that. Endogamy is what it is precisely because it is control over reproduction (by controlling people's sexual behavior, which must always have some sexuality and erotics in it); it is control over birth. Regulations on sexual relations regulate what kind of baby is born.

It is possible to argue here, then, that caste considerations come into operation even before a baby is born—only after its birth will the sex of a baby become a constitutive cognitive empirical difference.[1] In short, within these regulations, even the unborn already have a caste. It is assumed that it will be "of" the same caste as its parents. If the parents are not of the "same" caste, a baby's social position—even before it is born—is a matter of some anxiety. This, it seems to us, is what is truly remarkable about the difference between sex and caste. Although it is true that the sex of an unborn baby is of great importance in society, especially strongly patriarchal ones, nevertheless, barring sonograms and amniocentesis, it can be cognized only *after* a baby actually is born, and the sex is available for empirical identification: penis or vagina. To a large extent, the structuralists, including Freud and Lacan, were right: It is an issue of whether something is "on" or "off" (of course, they thought that only the penis was the "on" signal).

The caste of a baby, however, is determined even before it is conceived. Even if a couple does *not* have a baby, it is determined that if they did have a baby it will be of this or that caste. We have already emphasized that caste is predicated onto birth. Now it becomes clear why it has been extremely difficult to abolish it. We follow here a clue from Phule's first chapter in his *The Cultivator's Whip-Cord*, that the brahmans exploit a shudra-ati-shudra person from *before* he or she is born (or just conceived), and *after* the person is dead (because his or her progeny have to do something or other). We are transposing his argument against brahman ritual into a determinant category. It seems to us that like religion, caste is a determinant, rather than a combination of cognitive and determinant categories (as in the case of sex/gender). To put it differently, the sex of a baby has to be cognized first in order to become determinant, whereas the caste of an unborn baby is always already cognized and is always already determined. In the European history of this issue, its feudal history, the king and queen's *unborn* child would be a prince or a princess, almost exactly the same way the unborn child of a cobbler and his wife will be of the cobbler caste. Just as one's religion is determined before and after one's lifetime, one's caste too is determined before and after one's lifetime. One must ask: How can an unborn person have a caste? How can a dead person have a caste? They do. Caste, it seems, is not biodegradable. Derrida discusses the word in his "Biodegradables: Seven Diary Fragments," an essay we have mentioned earlier.

Also, one can wish to have a girl or a boy, but one cannot wish to have a baby with a different religion or a different caste. One can wish that the baby in the womb is male or female; one cannot wish that is Buddhist or Shinto. It is worth noting that what we recognize as "ideological" (following Marx), is, in the case of caste and religion, predetermined. In this context it is worth noting that there is a "category" for people of double-sex—transgender— but there is no category available within Indian thought on caste that might generate the category of "transcaste"; it only has "outcaste." We can only vaguely imagine what it might mean, but we believe the point is worth making.[2]

Various arguments can be made against the position we are taking here. It can be argued, for example, that if what we have called caste determinants are "cultural," then it follows that sexual determinants are "natural" (without being "biologistic"). It can be argued that the difference between the temporal sequence of determinant and cognitive categories is not very significant since the effect is equivalent. It can be argued that inasmuch as the religious and caste determinants are nonnatural, they are much more manipulable (transformable) than sexual determinants: It is demonstrable that it has been historically easier to break out of religious and caste determinations than sexual determinations, and this has been so precisely because religion and caste determinations have nothing really to do with the bare fact and practice of reproduction of bodies. The nonnegligible spread of atheism, and nonreligiosity, and the by now nonnegligible upward mobility of low-caste people proves this. Most of these arguments are valid in varying degrees.

However, inasmuch as religious and caste determinants, unlike sexual determinants, do not need to be anchored onto a body (an unborn, bodiless child has the same religion and caste as its parents), they can operate independently of human bodies. We should not be misled by the late-nineteenth-century and early twentieth-century attempts, by Europeans and Indians alike, to explain the "origin" of the "Arya" people and the racialist discourse it produced, supported by what we have called dubious forms of knowledge (theories of "race" supported by physical anthropology, for example). In a certain sense, these attempts at explaining caste in terms of racial typologies were looking for a way to anchor religious and caste determinants onto something tangible. This is paradoxical, because some of those tangible bodies had not yet begun to exist, had not even yet been conceived.

In that certain sense, this was the weakest moment for upper-caste intellectuals and indicates that the ground on which they thought they stood was already beginning to shake. Earlier, there was no need to explain caste at all; it was simply a natural order of human society, a fact of life and/or custom/tradition. At a certain point in the nineteenth century, brahman intellectuals felt the need to explain "caste" to the English and to themselves— the poignancy of the moment should not be lost. There is little evidence to believe that the English ever felt the need to explain "class" to Indians and/ or others. In any case, inasmuch as "class" was so thickly related to bloodlines in those days that it was more or less comprehensible to Indian brahman intellectuals, who had always believed in privilege derived from the simple fact of being born in a certain family.

These ideas seem to reinforce our earlier discussion, in chapter 3, of "societies of inheritance" and "societies of acquisition" (while constantly remembering that one never finds these in their pure forms). What we are attempting is a generalization, in terms of sociality, of the paradigm of "thrown-project" (in Heidegger's "analytic" of *Dasein* in *Being and Time*). Much more could be said here, avoiding metaphysics; for example, how Heidegger states that "*Dasein* is what we always are / I always am in each case." Our attempt here is to socialize, not that individual who should remain individual, but the notion that there is a certain unknown past and a certain unknown future. This is true of societies as well. The various distinctions that Heidegger makes among historiality, historicity, historicality could be made much more useful if we could use them for societies as well. We have attempted here to think sociably only about the "thrown-project." A preexistent past (known and knowable, subject only to operations of personal and public memory) and a preexistent future (unknown and unknowable)—faced with these uncertainties, societies seem to make a certain stylistic choice for the past or the future.[3]

Societies of inheritance depend much more on social memory and imitation (in the sense that Gabriel Tarde has given the word in his *Laws of*

Imitation of social distance or nearness that makes imitation of social behavior what it is), and societies of acquisition depend much more on "what is to come," on anticipation, on stealing beforehand, on speaking beforehand, on prediction. Consider, for a moment, the relationship between money, economics, and "futures markets": The function of predicting seems to have shifted completely from the prophet, or prophet-poet/artist to the economist, who seriously inhabits the prophetic function of predicting the future accurately, assisted by mathematical models mostly derived from game theory and by programmers who use probability theory.

Such a formulation might be thought unsupportable. If it is thought to be so, we could modify it a little to suggest that what we obtain are "styles or tendencies of sociality" (rather than forms of societies themselves). One style or tendency tends to look back, where "thrownness" determines much of the action; the other looks forward and the "project" determines much of the action in society. For the sake of completing the description we could also say that in the form of sociality that looks back, it is the memory, or even perhaps a rememoration, of custom that is dominant. We have already cautioned several times that these are often mixed and are not found in a pure form.

These are two forms of power as well, power in the Foucaultian sense, as "capillary distribution and circulation" with points of application that need to be identified (and are identifiable).[4] Further, theories and analyses of society, of power. and such like that take the metaphor of the human body to theorize about or to analyze society are entirely misleading. This observation is yet another benefit of giving up the model of power-as-sovereignty. It enables us to give up the metaphor of the body politic, both European and Indian (the purusha). It is tempting to look for another metaphor or concept for these forms, but perhaps it is wiser not to do so right away, for no single metaphor, however extensible, will be adequate for a good description of these forms.

To return to the issue of using the practices of untouchability as the markers of caste: It is important to do so because rules of intermarriage and inter-dining are found in other forms of discrimination as well. Using untouchability as the marker also has the advantage of taking the analyses of caste a little away from fixed ways of thinking about caste that have become entrenched in the social sciences. Inasmuch as untouchability operates over and above issues of endogamy, it is separable. It is necessary to see these as discrete phenomena (marriage rules and rules of eating together, and untouchability).

Since we are taking clues from the writings of authors such as Phule and Ambedkar and attempting to extrapolate a few of their ideas into a more contemporary study, it is worth considering what points of application of the caste system they thought could be identified. It is clear that Phule takes religious rituals as the point of application, and therefore resistance as well. The other is access to education, and through education to employment: A conspiracy of bureaucrats[5] prevents shudra-ati-shudra people's access to education and subsequent employment. It follows that Phule will construct another religion,[6] that he will appeal to the government for educational and economic attention. Ambedkar's writings are less easy to summarize in this manner, because he wrote much more than Phule did. We can see some basic patterns though. Because the caste system is the backbone of the Hindu religion, Ambedkar too will attempt a different religion. He too will insist on access to education and employment. Since he was in the position of a highly trained lawyer and was widely read and the chair of the committee that drafted the constitution of India, he was able to ensure access to education and employment. He must have seen that in a democratic state, the law must treat everyone equally; however, he makes sure that some people are given "special protection," thus marking the fact that those in power will not treat everybody equally at all. He also ensures that the state will intervene in what otherwise could be called civil society, usually not seen as a matter of state responsibility.

If our thought has to move away from the socioanthropological way of thinking about caste, we have to move away from kinship systems, purity/pollution, and above all, from the predication of caste onto birth. Therefore, in an attempt, if not to move away from but at least to turn away from the above, it seems important that we take practices of untouchability as the markers of caste practice. It becomes possible to imagine a way of seeing how these inscribe sociality with discrimination and inequality.

At this point it is important that we understand the nature of markers. These are to be distinguished from marks as such (markers always promise meaning; marks, by themselves, could be random and/or meaningless, and/or un-understandable).

Therefore, let us try and think what markers do. Obviously, they mark. The agentive is formed from the verb form, to mark. Or perhaps there is a threesome here: the noun "mark," the verb form derived from it, and then the agentive. The almost clichéd debate in ancient Sanskrit discussions of whether only the sentence means something by itself or whether the sentence meaning is derived from an aggregation or accumulation or serial comprehension of the meanings of separably individual words is partly relevant here. As is the debate from the same discourse on whether a noun is formed from the verb, or the verb formed from the noun: "Devdatta cooks." Yes, but is it not more proper to say, "Devdatta makes rice by cooking"? Does a verb result, always, in a noun, or vice versa?[7]

"I do." In Christian marriage rituals, this is uttered. Does it not produce a marriage? That is a noun. This is not a question about felicity conditions, as Austin wrote, and probably thought.[8] This is a question about the structural relationship between noun and verb. "Deodatus touches someone else," or, is it more logical to say, "By touching, Deodatus creates difference"? We should remember here the conceptual pair we discussed earlier, touching oneself/touching others.

Herschel,[9] Azizul Haque, and H. C. Bose, who refined Edward Henry's fingerprint classification system; Galton,[10] Faulds,[11] and others who discovered the uniqueness of fingerprints thought they had found a way of uniquely identifying an individual, because the whorls and curves and things like that are "unique" to the individual thus identified. "Unique" means there, in that discourse, that nobody else has the same pattern, therefore X is who X is, since the fingerprints of X match the fingerprints of X. This is as Heidegger in his *Identity and Difference* and, much before him, Frege in his "On *Sinn* and *Bedeutung*" had identified[12]—the Leibnizian principle of identity: A = A, we know this. Heidegger says this is a Leibnizian problem: "Identity" means "A = A." This notion of identity was improved—made more precise—by Frege (who had disagreed with Husserl's early work, and that had made Husserl change tack a bit), who asked, what does it mean, then, to state (proposition, judgment, statement, truth-claim, for which Meinong introduced a new notation, etc.), that "A = B," that the morning star is the "same" as the "evening star"? Or, "The morning star is the evening star," leading further to new developments in logic and language studies: Russell, Carnap, Wittgenstein, Austin, Grice, Searle, and Kripke.[13]

This technique of identification through fingerprints was also improved at the colonial criminology and police offices in Kolkata, and later in Pune. It seems that after the developments in crime detection and such other contexts as making contracts, checks, and so on. that somewhere along the line of historical development, signatures were institutionalized, meaning they became legal currency. The singular feature of identity, fingerprints, was increasingly relegated to criminology, especially with the rise of literacy (many more people could sign). This is an interesting historical development and needs an initial Foucaultian analysis, which we cannot undertake here. It is worth observing, however, that what began, at Henry's hands, as a signature for contracts was increasingly used by the state for identification and identity, as especially related to criminal evidence. Is it not strange, following Foucault, to see that identity is most easily correlated with criminality?

We already have observed, in chapters 3 and 4, that one always leaves traces on things that one touches. This "trace" evidence works fundamentally with the cause-effect metonymy, where the cause can be deduced from the effect: If our fingerprint is found on a glass, the conclusion is that we touched it. That we were present beside the glass to touch it is the inevitable conclusion (barring some clever criminal who might be able to "lift" and "plant" someone else's fingerprints, much more difficult than forging signatures). The presence of our fingerprints means our presence (assuming that the fingers are still attached to the hand, and the hand is attached to the body). That is why the state and the legal and policing systems can use fingerprints as evidence of presence.

It then follows that regulations on touching are regulations on presence and absence. Regulations on intermarriage and inter-dining seem to be higher-level phenomena, regulating social relations. Regulations on touching seem to operate at a lower level, regulating our presence to others and others' presence to us, and perhaps not so surprisingly, our presence to ourselves. Touching others and touching oneself, being touched by others, all these seem to be forms of self-presence as strong our own speaking voice. Untouchability then, in both its forms, seems to be a denial of bodily presence. It is not surprising at all therefore, that metonymies of bodily presence are submitted to similar regulations.

In this context, then, fingerprints and other traces of bodily presence (such as heat on a chair or a bed and body smells that are left behind) need to be understood in their *diffusion*. For these traces that are written onto the world are necessarily diffuse and in most cases unintentional. Regulations on touchability attempt to institute/ institutionalize these. Inasmuch as these traces are inevitable, such regulations seek to institute the category of choice onto these traces, and in many cases, they succeed. From this level, socioanthropological conceptions of purity/pollution and of inter-dining and intermarriage seem highly developed social forms that regulate relations among people. Regulations on touchability, however, work on both things and people. It is possible to say, "You cannot eat with X person"; it is possible to say, "You cannot marry X person." It is possible to say, "You cannot touch that thing/object"; it is possible to say, "You cannot eat that thing/object." However, it is not possible to say, "You cannot marry that thing/object." We cannot stop touching things as long as we are alive. This is the most diffuse and the most intimate form of life: touching things and people.

This allows us to suggest that regulations on touchability operate on the world and bodies in it with much wider range than regulations on eating and marriage. For example, before the nineteenth century, brahman musicians could not play most percussion instruments because they were usually made of leather, the skins of dead goats, cows, buffalo, and in some instances camel. Caste quite clearly relates to touchability here, and not to the music-making (the percussion instruments would be played by lower-caste people). This music-making group then might make great music together.

Lest we give the impression that in the discussion above we have taken into account only traces that we leave (deliberately or indeliberately), we should point out that the regulations operate also on traces that we might receive. That in fact is the greater anxiety behind the regulations: It is a bit like happening to step on a dead rat. Even though you are wearing shoes and socks, and there is no possibility of physical contact with the dead rat, that step leaves a trace on your left sole for a long time. This is not even a physical trace. Or it is like if you are a brahman and you are touched by a lower-caste person, or in earlier times, the shadow of a lower-caste person fell upon you. It will leave some trace, even if it is inadvertent. It is possible to imagine more pleasant encounters: Imagine you are leaning on a tree, looking at some birds in the sky, and the squirrels in the tree, chasing each other, run down your arm and leg. They are going to leave a trace in your memory for a very long time, perhaps lasting for years together, as they have, in our case.

Regulations on touchability then seem to attempt to regulate this diffusion, this multiplicity of traces. That which sociology calls social and anthropology calls anthropological is too limited, too small: Traces tell a different story. Traces cannot be narrativized; they are random; they merely cohere around our bodies in an act of remembrance. They can be united into a singularity only on the assumption that our body is a coordinated act. The heat of our body on the chair has nothing to do, really, with our fingerprints on the glass on the table, and the DNA from our saliva on its rim as we drink the water, even as the rustle of our clothes, the hiss of our breath vanishes in the ever-changing world of sound. It is this diffusion, this randomness, this unconnectedness, this passing away that made us forge the notion of the soul, something that will outlast the body.

Again we return to the Charvakin position: There is only the body, no "soul"; there is nothing paraworldly; there is no heaven or "beyond." In a certain sense, we have severed the assumed relation between religion and touchability/ untouchability, again indebted to the Charvakin position: Here is our body, and that's all there is.[14]

This notion of the traces we leave and receive was not available to the Charvakin thinkers. Neither was this notion available to Phule or Ambedkar—but they all had a feeling for these traces. Phule describes a shudra-ati-shudra household in the fourth chapter of A Cultivator's Whip-Cord, and it is clear that he has observed the traces and metonymies in some detail. Ambedkar, for his part, scholar that he was, trusts anthropometry and suggests that if a straight nose is some indication of anything, here is the nasal index, and so on (if these are the metonymies you trust, here are other metonymies). The trace in Ambedkar's writings is almost already a signature or a marker of some kind (what Phule sees as a moral/ethical problem, Ambedkar sees as an institutional/systemic problem).

Therefore, the ruptures they sought and created need to be scrutinized. Phule invents a new religion. Ambedkar repudiates Hinduism and embraces Buddhism of a modified kind, and many others follow him. The issue of institutionalization is much more serious with Ambedkar: After all, he *reinstitutionalized* the dalits in the most comprehensive and the most direct manner possible. There is something of interest here: His actions and his constitutional drafts and legislation did not make room for the dalits authentically to inhabit the margins, to remain locked up in what some people (dalit intellectuals) might misleadingly call their own "subalternity." Through his legal contributions he forced them to come out and be part of a democratic form of governance. We already have indicated the contradiction in the constitutional provisions, the special protection that he instituted. This eventually resulted in a self-consciousness that was equally contradictory, within a politic of identity: The dalits claim to be equal to everybody else and yet want to be treated specially, differently, with special rights that others cannot claim.

It might be a little exaggerated to state it in this manner, but the idea must be entertained. If in earlier times, untouchability was a certain institutionalized practice, the contributions by Phule and Ambedkar (and many others) reinstitutionalize dalits. It will not do, certainly not after these two intellectuals, to keep thinking of dalits as "excluded," and to think in terms of practices of "exclusion": That would be a traditional, brahmanical way of thinking. This becomes clear if we ask "excluded" from what?

It will not do, then, to think of dalits as "marginalized" or "excluded" or as "subaltern": They have been institutionalized within Indian society in a way that very few other groups were; over and above all the violence, all the breast-beating, over and above all the brouhaha over the Mandal Commission and whatever anybody, including me, might think or fancy as a caste problem.

This reinstitutionalization is a mark of a radical modernity that many of us still hope for. Without doubt these two thinkers were ahead of their times, perhaps even ahead of our own times. Nevertheless, it must be said that they did not have access to the ideas of the trace and the body that we have attempted to elaborate.

It seems to me that the traces we leave—even as markers of self-presence—are random, and much more stubborn and recalcitrant and obdurate toward any attempt at institutionalization than concepts of this or that kind of identity, that are usually predicated onto birth, inherited privilege, and other equivalents, including symbolic and economic capital.

It is only when we think of these two thinkers (and others like them) as those who reinstitutionalized dalits, that contemporary (1980s to September 2011) things begin to make sense in a dim sort of way: Statues, monuments, calendar art, pictures—what are these if not effects of an immense reinstitutionalization, one acknowledged (constitution, statues, public-funded celebrations) and another unacknowledged (calendar art, personal pride, etc.)? Is it surprising that Mayavati commissions immense monuments for Ambedkar and others, including herself? Is it surprising that scholars want to "work on," or "look at" or perhaps even "study" these phenomena? We have been warned of this "re": Derrida had warned us of "leaving" institutions (inside) only to reenact them in a different way (outside).[15]

Therefore, the need arises to ask the ancient question again: What are the markers of caste practices? Practices of touchability and untouchability are these markers, and not merely regulations on legitimate sexual relations, marriage, or eating together. The latter, regulations on sexual relations and so on, are far more institutionalized than the former practices of touchability/untouchability. In fact, precisely because of the randomness of the former (which generates far more anxiety than regulations do) they cannot be fully institutionalized or reinstitutionalized. This might be one explanation why touchability and untouchability have not gone away. This might be one explanation of why Ambedkar's insistence on intercaste marriage has disappeared from contemporary caste politics, such as they are.

Thus, it is possible to generate a classical structuralist ratio (we cannot go beyond structuralism without passing through it in some way): Fingerprints (metonymies of bodily presence) are to touchability/untouchability what signatures are to intermarriage and inter-dining.

Fingerprints : touchability/untouchability :: signatures : intermarriage and inter-dining.

Structuralism of the Lévi-Strauss variety: There are eight hundred myths to be understood, analyzed, interpreted: Something abstract is needed to "comprehend" that abundant multitude. When you find a pattern, you have got to have some notation for it, some way of writing it out.[16] When you reach out and get it, you do or do not realize that it is Aristotle's description of metaphor: A : B :: C : D (the shield is to Achilles is what the cup is to Ares). It is clear that in this ratio the abstractions A-B-C-D are blank slots that are, in this particular instance, occupied by "shield-Achilles-cup-Ares." It is because it is a ratio that Aristotle can say that the A of D means the C of D, and the C of B means the A of B: the shield of Ares means the cup of Ares, and the cup of Achilles means the shield of Achilles. Metaphor then seems to assume the metonymy of invariable concomitance (always happening together). It has to be noted that this was a ratio, not really an equation, and not a formula (though most of us have used this ratio as if it were an equation or even a pair of "binaries," for exceptional analyses of social phenomena by people who must have known some mathematics).

What we are attempting to articulate is the thought that practices of touchability/untouchability operate below the level of intermarriage and inter-dining, and because acts of touching or not touching are far more frequent and far more random (in the sense of unpredictable), they have never been regulable. We include here the accidental but scathingly erotic touch of a table, or a cousin or some other person, a cushion perhaps, or the door handle or tap if it matters or, as always, the painful cut from the knife accompanied by onion smell and tears, distraction and randomness; and therefore, there was an attempt to regulate these acts. There are all manner of things attached to these acts, there are Fregean meanings and significances; we could merely say *Sinn* and *Bedeutung*, or in a literary way, literal and figural, denotative and connotative, judgment and quasi-judgment and so on and so on, including ancient Sanskrit "rhetoric." It always has been a question of what it means, and what else it means. How do we distinguish between "literal" and "figural" meaning? How do you know that when I say I, I may not, on second thoughts, in fact do not, mean me? That is the difficult "we."

After that brief *divertissement* let us return to what really is the issue: Since our bodily presence leaves traces, and since other persons and other things leave their traces on us, we need to find a way to understand (regulative function of reason) these, and we could and can only understand through schemas. The elaboration of singularities is a never-ending task that leaves us without understanding.

So those who did not want to be touched by certain kinds of people ended up with a schema, which in history became a scheme and later a framework, and even later led to a social practice and custom, ending up as "tradition" (self-legitimizing). Perhaps it's not a remarkable scheme, since it merely states and performs the *Sinn* and *Bedeutung* of the statement "I shall not be touched by X." Now it is clear that denial of this kind, or even a refusal entails certain actions, of commission and omission. If we do not want to be touched by X, we could run away (as women often have to), or we could forbid X from touching us (as women usually cannot). It's anybody's guess as to what happened in history that it could harden so. We think this was the basis, touchability/untouchability, and not regulations on intermarriage and inter-dining. The latter are derivative phenomena, merely social in the developed sense of relating to people (including oneself). Our bodies however, come into contact with things as well, and do so every moment of our life. Regulations on such contact establishes the custom of touchability/untouchability. For this reason, conclusively, we think, that touchability/untouchability is the marker of caste practices.

TRANSPOSITIONS: THINGS AND PEOPLE

To put it simply: People who touch things that we do not touch *become* untouchable. Our unrelation (we need this awkward coinage because it's an action of not relating deliberately; it's not an indifferent nonrelation), our denial to relate to certain things is transposed onto people who do touch these things; perhaps superimposed is a better verb to describe the phenomenon. This is quite clearly, again, an operation of metonymy, with a sort of inmixing of the two basic kinds: invariable concomitance (part-whole and frequent proximity), and cause-effect (a temporal invariable concomitance). Those who do not touch animal fat, skin, human waste, and so on will not touch people who do precisely that, because of their frequent contact with these things.

We need to understand how this happens. Invariable concomitance and part-whole relations function in a certain way where inanimate things are concerned, and in another way where animate human beings are concerned (we will refrain from complicating the issue by introducing our relations with animals). By things we mainly mean artificial things, things of human production. Of artificial objects that are called art objects, we shall not speak here. We wish to focus on things that we use for physical reasons. This would include the hair-trimmer, the hair-remover, airplanes, the European TGV trains, the Japanese bullet-trains, the coffee machines, and what have you, including the polymer-science product of the plastic water bottle, Coke, beer, rum, whiskey, a lot of food that is prefabricated, and so on.

Though it is inevitably tempting, we will not give in here to a Marxian description. If we had wanted, we could already have used the term *reification* and written that relations between things are confused with relations between people. That however, is not the case, because we have here metonymies to deal with, not analogies, certainly not metaphors or allegories.

We are not writing here about relations between things. We are attempting to understand how our relations with things are *extended* through metonymies to include people—because *in principle* it should have been impossible.

To return to metonymy then (the part-whole metonymy and the cause-effect metonymy), the former depends on a spatial concomitance, whereas the latter depends on a temporal concomitance. We have already pointed out that metonymy often looks like logical inference—this is particularly true of the cause-effect metonymy (which in many cases *is* in fact a logical inference). What we are attempting is to point at the intellectual shift between our relations with things and our relations with people.

In our own literary way, this is a question of the difference between the cause-effect metonymy and a logical inference. A cause-effect metonymy (there's smoke, so there must be fire, since they are most often found together, temporal and spatial invariable concomitance) is almost indistinguishable from an inference (there's smoke, *therefore* there is fire).[17] The part-whole metonymy depends on the spatial invariable concomitance, although it is a little more complicated.

Let us keep that line from ancient Greek thought, the Aristotle in Descartes, and the Aristotle in Kant and Marx, and later the Descartes in Husserl. This is the matter of *extension*: How do we extend "I don't touch shit" to "I don't touch those who touch shit"? We have attempted to explain this by calling this process "metonymy." That does not explain what makes it possible.

Let us assume that our relations with things are different from our relations with people. We manipulate things by touching them, taking them apart, putting them back together. Things are inanimate (these are metaphysical assertions, but we see no way to avoid them), whereas people are not. We know how things work (since we made them work in that particular way in the first place), whereas we do not know how people work ("what makes you tick?"). Things are culturally produced; humans are produced through sex in a sexually dimorphic species: reproductive heteronormativity.

Artisanal production of things (again, thinking of Simondon's essay mentioned earlier) is very often thought of as if it were parthenogenetic—a single person makes a small tool, chipping away at the stone. A single person paints a canvas. A single person writes a novel. The other person is *not* party to the physical/intellectual production of the object. A woman or a man cooks food in the kitchen without any "outside" help, assistance. A painter works on a painting for several months, and nobody knows what he or she has been doing. It seems to me that such solitude of production—the Romantic idealistic description of "creativity"—is a dream of parthenogenesis of a dimorphic species. That is a dream, for there is never a moment when one is solitary, nonsocial.

The truth of that dream now, however, is different. It is very rarely that one makes one's own tools now. Tools too are very often mass-produced within industrial capitalism, a mode in which production can be modeled neither on parthenogenesis, nor on dimorphism. It can only be called *polymorphic social* in this context. Therefore, it is possible to argue that our relationship with things is polymorphic social not only in terms of production, but also in terms of circulation and consumption.

We can safely assume that we touch things more than we touch human beings and we touch artificial/inorganic things more than natural/organic things. It is this dispersed hapticity, this haptic dispersal as it were, that defines the life of our bodies as much as anything else. The polymorphy of our sociality becomes visible when we focus on *how* we touch things: There are uncountable forms in which we do that. It also becomes clear that precisely because this polymorphic social touching is dispersed and cannot be regulated that it needed regulation.

It is clear that there are more regulations on touching people than on touching things. We have already established earlier that things that we touch most are things that we "own." It is possible that such "possession" is transposed onto people as well. The patriarchal possession of women (brought out so clearly in the 1971 film *Ceremony* by Nagisa Oshima) is one example; there must be many others. It is through such repeatedly touchable ownership that we might get a glimpse of what legal possession is.

Perforce we have to attempt to imagine historically (we do not know any other way to access history and historicity). It seems to me that in the nomadic/hunting-gathering mode, there would be *very* few regulations on touching things or people. We have to be very careful in imagining that condition: There are no marriages; there are no families; there is no possession of anything; there is no electricity, no guns. There is very little individualization: Everything one does is in the service of group survival or for one's own pleasure (individuation *without* individualization). However, one thing that we had asserted seems not to be true in this particular situation: There must be things that one touches frequently, but these are not one's own *possessions*. There is singularity, but there is no individualization of objects or persons. There is no need to clean up after defecation. There is no need for "dirty work." Everybody is eating whatever is edible, including animal entrails. As Marshall Sahlins has shown in his *Stone Age Economics*,[18] the work cycle is dependent on the body cycle and on the earth's diurnal course.

Also, if we pass touchability/untouchability through the Marxian grid of the five or six modes of production, we might get other answers. Regulation of touchability/untouchability must have come into practice in the "agrarian" mode of production, when ownership of pieces of land began to territorialize everyday living, and private property came into being. These forms of property became rigid custom in the feudal mode of production, and in the capitalistic mode of production (which might have deterritorialized some aspects of everyday living and making things), they were modified. Naturally, this is a loose description, since we might not obtain the historical linearity associated with these transitions: The regulations on touch therefore are sometimes loose, sometime rigid. We have already pointed out in chapter 3 how in a crowded bus or train the sense of touch is willfully "benumbed."

Thus, it seems to me that regulations on touching things were transposed—through convenient metonymies—onto people. There are actually two processes: the thingification of humans and the humanization of things. These two processes enhance and extend and confuse regulations on touching and being touched by people and things.

With our focus on caste practices, we need also to point out here that in the realm of production we touch things in order to act upon them in some way. We can observe that historically the changes that have taken place in the technologies of production are from artisanal production (unmediated use of the tool directly by hands, feet, and teeth, etc.) to robotized/automated production where the motor skills of hands and feet are used to control machines that act recursively upon machine that act upon things to make some other thing. Thus, it becomes possible for a pack of edible items to bear the description "untouched by human hands," "purely automated production," etc. We take this up in the next section.

To return to things and people then, we can observe that there are transpositions and transplantations taking place between regulations on touching and being touched by people and things. Some things might become untouchable because they are associated with untouchable people, and some people might become untouchable because they are associated with untouchable things. Although many such associations are "arbitrary," some at least seem to have their bases in operations of metonymy.

TOUCHING TO MAKE AND TOUCHING TO PRODUCE

A history of touching and being touched in the context of making things is necessary. Although we cannot even begin to undertake such a daunting task here, we will observe general tendencies, taking clues from Marx, Simondon, and others.

In prehistoric times, when things—tools, clothes, decorative marks, etc.—were made, the maker used his or her hands and "manipulated" some things to act upon other things. Minimally, hands were used without tools, for example, to mix clay with water to bake a pot. This touching was physically unmediated. The tool, when it was used, mediated between hands and the thing manipulated. It also made a degree of precision and a degree of repetition possible. Later, machines were assembled, and people touched the controls on machines—relays—in order to make things. The machine could assure a much higher degree of precision and repetition. The machine could, in principle, serially produce objects en masse. With the use of mathematics and geometry, the machines were made even more precise, and could now produce identical objects en masse. The controls on machines too were made more precise, though they still were analogical controls, and depended, in the final instance, on the motor control of hands and feet. Skilled labor consisted of knowledge of the machines and the delicate working of controls. Unskilled labor consisted of not-so-delicate manual operations—tightening screws, fueling the machines, and so on.

Even later several machines were articulated onto each other to give rise to the industrial factory, the weaving mill. New forms of energy were discovered—petroleum and electricity. With the coming of electricity things took a new turn, for unlike coal and petroleum, which had merely to be extracted from the earth, electricity itself had to be produced: The energy itself had to be produced. This renders the processes of production more complex. One has to use naturally found energy to produce artificial energy, and to unearth naturally found energy we can use artificially produced energy. It is worth asking the question: What form of energy is used in nuclear power plants to produce electricity? The answer looks simple: atomic energy. But to release that energy from atoms of specific substances requires energy, which energy must be produced; to produce that energy we must unearth coal and petroleum; to unearth that we have to use electricity or steam, which in turn must be produced. It looks like an infinite regress, but it is not one, because the baseline is that energy is produced using coal or petroleum. It seems that we can safely conclude that mining is the basic industry (extracting petroleum from earth or sea is a form of mining). In each one of these processes some energy is lost, dissipated: entropy.

The implications of Simondon's observation that the tool is operated by human energy directly applied onto the tool and thing to be made whereas the machine is operated by artificial energy begin to be clear now. It seems to me that for the previous, we could use the verb "to make," and for the latter we could use the verb "to produce." This is necessary because the training to make things is usually elder to young, master to apprentice, teacher to student. Whereas the training to produce can be learned in an educational institution, or "hands on," mainly because it is less of learning to make than to use machines correctly. Also, in making things, one learns by trial and error, whereas machines pro-duce (lead beforehand) according to a well-formulated plan that is later on, by operating the machine's controls, executed.

Later on, the operation of the machine's controls is written into an algorithm, a "program," a marking beforehand. Digitization in the latter half of the last century has been overvalued—it has often been touted as the latest revolution and such words as virtual and *simulation* have acquired a power of their own, as if they were autonomous realms independent of industrial production, independent of mining. But the various alloys and polymer syntheses that are used to house the circuitry on a chip are produced industrially: The digital rides on the shoulders of the industrial.

Also, the tool was much less determined than the machine. The tool could be used for some other purpose, as a tool. Anything can be used as a tool, depending on its material properties. But the tool is a special case of usability, because it has usability that we inscribe on it: We shape a stone, give it a point, after which it ceases to be just any odd usable thing and becomes a tool. Machines are much more determined. It is more difficult to use machines for any purpose other than the one they were assembled for. To put it quickly and simply, one can use a key to turn a screw, lacking a screw-driver, but one can use a blender only to blend. We could imagine a horrific scene where a murderer cuts up the victim and patiently blends organs to dispose of them; however, the action performed is blending. The contents of the blender are unusual, not the action of the blender. It is worth remembering a line from the second paragraph of Marx's *Capital*: "To discover the various uses of things is the work of history." In that sense, machines tend toward an a-historicity. This is evident when we revive some nineteenth-century machine and make it work (some examples can be found in technology museums, for example).

Not very long after, it would seem, we find ourselves surrounded by machines articulated onto other machines in an inexhaustible network of mechanized mass production of innumerable commodities. Sometimes, this is called "technology." It is clear that technology is a condition within which we find ourselves, not this ensemble of machines, whether digital or analogical. It is possible to think in terms of three or four layers, organized hierarchically: At the base are hands and feet and their motility. This is superimposed by tools. These in turn are superimposed by machines and mechanization. The last layer is "technology": a condition within which we find ourselves. Using these layers as a heuristic, it should be possible to chart various cultures in terms of technology without giving in to the default and therefore usually absent adjective "advanced."

Such a historical charting would also be the history of touching to make and touching to produce. It is to be observed that touching to make is more or less unmediated touching, whereas touching to produce, at the current conjuncture means *not* touching the thing to be produced—it means touching the controls of a machine, and increasingly, tapping away at a qwerty keyboard. After that the mediating machines take over the task of production. It is no longer necessary, almost, to get one's hand dirtied, sullied, stained, bloodied, or greased with animal fat.

Mass production produces identicality and identity as well. It is necessary to keep the distance intact between identity and singularity: In mass production of commodities singularity is a flaw. This dent on this car that we are about to buy is singular—no other car has that exact dent—that is a flaw. We will not buy it because it has that dent.

In making things, the imprint of the maker's hand is usually visible (recalling Benjamin's essay "The Story-teller"), and that is what added value to the singular thing. It is exactly the other way around in mass production which does not countenance, let alone tolerate, singularities.

I have six toes on my left foot, and no mass-produced footwear is usable. It becomes clear that in making things, let us say shoes, I could take into account the six toes that my buyer has. I ask him to put his foot on a piece of paper, draw the shape of his foot on it with a pencil. Then I cut leather from that stencil and wrap a shoe around the six toes. I take the measure of my buyer's torso and produce a shirt with those measurements. I do a temporary stitch and ask him or her to wear it. Then I put a pin here and a pin there to make it fit properly, without constricting his or her movement of arms and chest and neck and pectoral and trapezoid muscles. Then I finalize it. The entire process is negotiable. If the buyer wants something stupid, that is, something that cannot be done, I persuade him or her of how things are. We are not talking here of sentimental "human" contact (although that too is what is going away). We are talking here of the nature of making things: There was always some improvisation necessary. Something "unforeseen" (im-pro-visation) could always take place. This area of improvisation partly overlaps with bricolage. The relationship between bricolage and the line that we have quoted from the second paragraph of Marx's *Capital* is inferable.

With that, let us return to our basic concern: What is the difference between touching to make and touching to produce?

A woman turns the wheel with the stick, and her husband shapes wet clay with his hands. The clay is in contact with his hands. In fact, it is essential to make a pot that he controls the dripping clay with his sense of touch, coordinated with other senses. He hears the khadak-khadak-khadak-khadak of the wooden wheel as it turns; he sees it turn at a certain rate. He slows down when it slows down. She turns the wheel again. This wheel, in case you do not know, is horizontal. This is not a metaphorical wheel of fortune, neither is this a car. This is the potter's wheel, and in Maharashtra at least, only the wife of the potter can turn it. At the center is a heap of clay that will in some time become a pot. The daughters are looking on curiously; the sons are fighting over a piece of wood. The woman puts the stick onto the wheel and gyrates it. It turns on and on. The husband works faster and forms the beginning of a rim. The rest of the story follows. It's a nameable sequence, à la Barthes: baking a pot. You put water in clay to make it manipulable, and when the pot is formed, all the water is more or less gone;

you pick up the pot and put it in the kiln and put heat into it and bake it. You have just made a pot.

With touching to produce digitally, the story is very different. All she touches is the qwerty keyboard[19] in a certain "language": Basic, BasicA, COBOL, FORTRAN, C, Visual Basic, C++, C#, LISP, Scheme, Python, Perl—there are hundreds of programming languages—and in the beginning of it all, Assembly, what is so happily, almost proudly called "machine-readable." That last word actually means executable: Having received a certain electrical input, machine parts will move in a previsualized, beforehand-marked process (knowledge of action before action) and produce one mass-produced commodity at a time. To be uncouth: It is time someone, *anyone*, analyzed such programmatologie.

7

SOCIETY, SOCIALITY, SOCIABILITY

SOCIETY AND SOCIABILITY

It would be the gravest error to assume that there is something, one thing, that we could call "society." It is precisely the lack of a society that we have been attempting to get at. "Society," derived from *socius* (companion, follower, etc.), is a European, indeed an ancient Roman, notion. If the ancient Sanskrit descriptions of the fourfold division have some truth (in the sense of phenomenal social content) in them, it would mean that there were four societies. This is neither an issue of plurality or multiplicity nor of undecidability: This is an issue of whether there is or is not something called society. It would not do to seek an alibi in the above-mentioned notions. It is time that the bluff was called.

We have already seen at least twice, that the word that Marathi uses for society, *samaj*, also means caste, clan, community. So there are, then, many of what is called *samaj*. This is not a happy plurality—it is segmentationalist. It is tempting to suggest that it is segregationist. (We are suggesting that it is segmentationalist [there is a cut] rather than segregationist [gregarious only among themselves] because of the century-long and complex attempt to establish an analogy among, if not to equate, race and caste and segregation.) There is an intimate cut: We interact with but will not relate to that other *samaj*. The members of that *samaj* are not from ours.

It becomes clear, though a little darkly, that there never was "a society," not here and now, in Pune at 13:50 on April 14, 2012, nor in equally historical Europe, and equally, not there in Rwanda, not there in the no longer there Biafra, not there in Cambodia of the Khmer Rouge, not there in Ahmedabad of the Gujarat Riots, not there in metropolitan malls, not there in universities, not there anywhere—not in the metropolises, urban centers, rural villages. That a large number of people display regularities in daily conduct and behavior and relationships does not necessarily mean that they belong to the same society, does not allow us to conclude that there *is* a society, or that there *are*, in fact, social institutions.

What can be found is *sociability*.[1] It is (has to be) invented: Every moment of interacting with those people we do not know we have to find that moment's sociability, à la Emily Dickinson who wrote, in a poem, "The soul selects her own society / then shuts the door."[2] That is what *samaj* means. Family is customary space, as is our own professional space, classroom, programming cubicle, gym, bus, local train; it is all familiar. But when we have to interact with those of whom we have no knowledge whatsoever? Sociability—as we see it now—is a matter of how one relates to "others"—those persons and things that we do not know at all. This is partly an issue about alterity and partly about sociability and, shall we say, a sociable encounter with alterity?

One European solution was to respect the rights of others (because one thought one had rights; therefore, others too must have them), so that others respect our rights in turn; a certain reciprocity is assumed (never fully proved). This is where the problem crystallizes itself. This is about the so-called idea/l of democracy: We are all equal and therefore must be treated equally.

Even there, the notion that we share the same phenomenal space, the same phenomenal, material world, is missing, since it's a matter a nonphenomenal, impersonal, socially necessary abstract equality.

Driving to our workplace every day for a variable forty minutes and back in another equally variable forty, traffic (we know we are also traffic) has become the symptom of a pathologically nonchalant nonsociability (we are slowly getting rid of the word *society*).

The road could be conceived as a space that all vehicles share—because the municipal nature of our cultivated urban life built the road—the vehicles let each other pass, signal, slow down for others when they fumble the clutch and the car stalls and so on. That, however, is not what we obtain: The nonmunicipal philosophy of driving in Pune (bike or car, bicycles no longer dare have such a philosophy) is: "The road belongs to me; the others are obstacles."

There are no other people going to work, or home, or wherever. They only are obstacles that we have to overtake, circumvent: We turn whichever way to overtake and be there first at the next signal, to be there home or workplace first. In brief, others only are Newtonian bodies. They have their dynamics, which we are perfectly able to calculate and drive around without even nudging anyone. We are all good drivers. However, everything except me are just physical bodies that obey regularities that Newton identified, and made into verifiable, and repeatable and iterable calculations. We are human, perhaps all too human, but others are just obstacles, not even adversaries, not even competitors for space and territory. They become that—adversaries and so on—only if they brush our car or our body. Then it is touch and go, and the image of "road rage" popularized by the media, seems to explain everything. Urban legend has it that one can get killed while driving—that in Delhi people pack guns and use them to kill someone who obstructs the projected acceleration and trajectory of their car. In Mumbai too. In Pune, a bus driver goes on a rampage, marauding his way, safe in the bus, through morning traffic, for an hour and half, killing about thirty people. This was called "extreme road rage" in the newspapers.

One would think that a road, a shared convenience, could not ever be "territory." But we have just managed to make it into our territory. There is no society. There only are forms of sociability, some decent, some indecent, some aggressive, some not, some neither here nor there, and worse—some so narcissistically nonchalant that they are altogether indifferent to sociability—unsociable. Let us try to be accurate: As stated earlier, with much conviction, things that we touch most often with a minimum expenditure of psychophysical energy are things that we possess, things that belong to us. "Things" that we have access to whenever we want (we are aware of the fact of marital rape and what such access might mean to women raped thus). The easiest touch is when we touch ourselves without knowing that we are doing so. That is when we are most proximate to ourselves, without our knowledge, but with our sensation.

One touches oneself all the time. Our senses—driven, no doubt, by our body image—tell us that we are doing it. But we do not know it. This is particularly evident in Pune, where men scratch their scrotums, fully clothed, of course, and they do not know or even perceive that they are doing it. That gesture, studies of the great apes tell us, is a gesture of aggression: Chimpanzees and gorillas do it to signal some possible danger (that is when aggression kicks in: It is always somebody else who made me do it).

Let us return to society and sociability. There are the streets with their vehicles, with their jaywalkers, with their conversationalists, hawkers, lines with shops and houses and offices: streets and buildings. ("Look at all those lonely people; where do they all come from? Where do they all belong?") What are they doing?

A description that now sounds alien to us is: They are individual members of a society interacting (or not) with other individual members of society. Another description is: These people are recognizing themselves in one another (or should we say, "each other" and reduce this matter to a manageable Hegelian pair again? But if a third comes about there, why not a fourth and fifth?). There are many descriptions, including "being-in-the-world" and "being-with." Perhaps this was the phenomenological content of that description, that sociability has to be invented every moment.[3]

What was called "society," and the various "social institutions" that sociology studied and partly invented out of its own internal disciplinary requirements, were customs and forms that made it easier to invent sociability: forms, not phenomena, certainly not "social phenomena." These customs and forms were more or less rigid and flexible at the same time. They were, however, always easy to lose: That might explain why violence breaks out so easily when we are unable to invent a customary or a new form of sociability when these customs and forms fail "us" (or "them"). It is the very fragility of custom that makes rabid assertions and violent enforcements of it so necessary.

The "social" nature of "society" becomes clear when we ask ourselves, how many people do we really know? A very generous estimate would range between ten and one hundred—in the present. We do not have total recall and may not always remember intimate or nonintimate moments with someone other. That is the limit of our understanding of society: The rest are all unknown human beings. Therefore, the need to distinguish between society and sociability; the latter is about how we conduct ourselves in relation to persons unknown.

There is another way of approaching the issue that we are trying to make visible. This approach is via performance art, especially that of the German artist Stefanie Trojan. Several, if not most, of her performances take place in such public places as streets, and especially museums and art galleries. Several, if not most, of her performances involve going up to strangers and doing "pleasant" or "unpleasant" things: touching them, jumping up and down in joy in front of them, picking a speck of dirt from their cheek (grandma's spit), carrying a portable toilet and sitting on it in public with pants down; standing half-naked with just a jacket on, in front of a painting, grabbing and eating food from someone else's plate at a party, lying down in an art gallery and holding people's legs and not letting go, posing nude beside a sculpture of a nude, standing half naked beside a clothes rack, asking people to dress her. A good list and videos of these are available on her website. In several, if not most, of these, she does not speak.[4]

Our first glimpse into the issue we are trying to make visible came from somewhere else, from traffic and loving couples in public, unaware of where they were. However, Trojan's performances crystallized things for us a few months after we had seen one of them and had seen others on her website. Our first interpretation was that she was trying to convey something about language (because she does not speak). It followed that she usually explores nonlinguistic sociability. But her nude and half-nude and "dirty" performances seem to explore sociability itself. Initially, we thought she was exploring and testing social norms.

But when we attempted to understand the various "reactions" of the people that she "subjected" to her performance, we realized that she was attempting to get strangers "to formulate" or "to improvise" some—any—form of sociability, and that too at that moment itself, in an interaction with another human stranger, to perform some sociability. The recipients of her performance had to perform something "social" themselves, while being individuated.

This alerted us, though it took a long time, to the possibility of understanding human interactions in terms of sociability instead of society and institutions within it. From then on it became clear that most study of society—most sociology—took for granted a relatively stable object of study, revealing the "positivist" nature of most sociology and anthropology: not just Comte, but also Malinowski, Lévi-Strauss (though he came almost away from these disciplines), and perhaps even Bourdieu ("habitus"). This was the "science" part of the "social sciences," most study requires, by its own internal dynamics (nature), that the object of study be relatively stable.

We had a glimpse of this problem when we suggested earlier, in chapter 4, that there seem to be very few "sociological" studies of "temporary communality" (that was a tentative and incorrect formulation): There are sociological studies of "suicide" that are nearly archetypal, but there is no sociology, to the best of our knowledge, of the girl's falling body as it falls while she, high on this or that form of joy assisted by acidic chemicals, thinks she is flying; there is no sociology of her joy as she flies before she splatters her blood and brains and bones six stories below.

Similarly, there is very little sociology (this does not mean there is none) of the "flash" crowd and its violent or pacific behavior. It seems to me, again, that Gabriel Tarde's notion of imitation could be used here to undertake a different kind of sociology altogether. For it is not just question of flash formation of the crowd, it is equally a matter of its flash dissipation. In a pacific flash crowd, it is the moment of dissipation that is most fragile, since the dissipation of the crowd also might come to mean the dissipation of its pacifism.

SPATIAL AND TEMPORAL SEGMENTATION
AND REARTICULATION

The relative stability of the object of study brings up the question of spatiality and temporality: "relatively stable" in time and/or in space? We have been aware that for the manner in which we have proceeded to think until this chapter, temporality has been an unacknowledged embarrassment (inasmuch as we have not been able, until now, to ask the question "Does caste have temporality? If yes, what is its temporal nature? If no, why?"

The common vocabulary of "higher" and "lower" and "forward" and "backward" gives us a significant clue—it fuses together spatial and temporal metaphors. It is clear that the spatial metaphor indicated a belief in a static hierarchical social organization. It would not be too much to suggest that what the anti-caste struggles have achieved until now was a radical modification of a spatial metaphor into a temporal one.

The "lower" castes were now rearticulated as "backward," and that too in terms of "progress," "modernization" "equality" and "upward" mobility. (We are aware of how "forward/ backward" are themselves spatial orientations; however, these are orientations of anticipated/ remembered [protention and retention] movement, and movement itself is temporal; therefore, these could be, and are, used as temporal metaphors: forward meaning future, backward meaning past, a metaphor of a body moving and perceiving and remembering.) Until now the anti-caste struggles have remained within the domain and paradigm of "modernity," almost always understood in terms of capitalistic development and access to economic and cultural resources. The point we made about Phule and Ambedkar in chapter 6, that it should be possible to understand them as thinkers and actors who *reinstitutionalized* dalits into society in a new way resonates now.

This also allows us to understand the strange (to us) common belief in the unchanging nature of caste. It was possible to see caste as unchanging in time because the dominant perceptual metaphor was spatial—and within the Hindu metaphysics it was impossible to find a way of using temporal metaphors that had social change and "progress" at their metaphorical heart. Such changes as inevitably must have taken place were rendered invisible or negligible by the metaphor of a hierarchy that was always heuristically available, ready to be invoked at the slightest provocation, and even more strongly, always seeming to make sense of the existing social organization. It is almost as if this metaphorical structure were the condition of possibility for something like caste to be experienced. We remember here Kant's notion that conditions of possibility for the existence of something are also the conditions of possibility for the experience of that something (something similar to what in the physics of the 1990s used to be called "the anthropic principle"). We have indicated earlier how this metaphorical structure guides everyday actions of people. For most people, "caste" has a cognitive force, the force of cognitive stereotypes. What Phule started and Ambedkar continued and completed could now be seen as the first *rearticulation* of this spatial metaphorical structure. Ambedkar, more than Phule, rearticulated it into a temporal metaphor of "backwardness." On the surface, it looks as if nothing much changed because of this rearticulation. "Upper" meant "forward" in social practice ("upper" classes were also the most "forward" it seems), and "lower" meant "backward." However, since progress and social change are at the heart of these terms, they are not actually symmetrical: "Backward" is what it is, not just in comparison to "upper" classes who might think themselves forward—it also compares its own present state with its own future state, as *imaginable* in the present.[5]

Subtly, almost imperceptibly, almost unconsciously, something had changed. The *spatial* metaphor had been *temporalized*. This is neither to suggest that comparisons with the upper castes were not taking place nor that the "backward" did not imitate the "upper." We suggest that even if some imitation of the "upper" castes was taking place, this was under the sign of the future. The revengeful ressentiment that can often be seen in dalit psychopathology is generated, it seems to me, by the (perceived) lack of material and cultural access to that projected future of equality. This subtle change has been neglected by much of the thought on caste—especially because much of the political activist thought has depended on a supratemporal and suprahistorical notion of caste identity. The temporalization opens up more comparisons, not just the spatial-hierarchical one. It opens up a comparison of the pasts, the presents, and the futures of the various castes. That which was an a-temporal hierarchy is now segmented into the temporality of past-present-future.

It is here that the "modern" idea of time as open-ended and of the future as unknowable but workable begins to take root in society. We are not suggesting that Phule and Ambedkar were the ones to do this. This had already happened, quite clearly, when the changes in the movement of capital deprived people of their inherited occupations. They had to do something in the present, in the face of the bleak future they could easily see for themselves. Phule and Ambedkar institutionalized the *workability of the future,* the idea that I can change what seems to be my future by doing something in the present. They also made possible the idea that the future can be worked on collectively.

People were disinherited—had to disinherit themselves—and be open to some unknown future (for daily sustenance). This is a complex moment, since in terms of psychopathology, this disinheritance can be disabling and in some cases traumatic. It could also bring about a nostalgia for skills and lifestyles that are no longer usable and workable. The socially acquired fictions/ cognitive metaphors did not work anymore, in the face of economic changes (which seemed to threaten "survival" itself). In Marxist terms, the content had run ahead; the forms were lagging behind.

At one point in chapter 6, we almost had hit upon this point when we tried to indicate/ believe that we are trying to think of Heidegger's notions of historicity, historiality, historicality, and so on in more explicit social terms. At another point in the same chapter, we failed to see the connection between his notion of "thrown-ness" (as we understand it) and our own emphasis on "inheritance." If the notion of the "thrown-ness" had a certain "high modernist" sadness to it, the notion of inheritance adds a penumbra of joy to the existent darkness. The "project" aspect of "thrown-project" still retained thrown-ness despite the prefix "pro" because the "ject" was emphasized.

It might seem as if we have got distracted from segmentation. However, as Saussure was clearly able to show, segmentation is a temporal issue as well. When one listens to a stream of sounds coming from someone, one is able, clearly, to segment words from that stream of sounds, and syllables in a word, if one knows the language that was being used. This is the magic that human beings perform when they know the language. It is like seeing a face in a painting, or a drawing. One is prevented from seeing patches of color or a sequence of lines.

Thus, cryptically again, we might be able to say that segmentation is both spatial and temporal. Our bête noir has been, for some time, the "organization" of spaces, and the relationship/s or nonrelationship/s in them. We are suggesting an analogy between newspapers and housing. Just as the news on the left has nothing to do with the news on the right—in a newspaper— the neighbors on our left have nothing to do with the neighbors on our right, and we are not in the middle either, because we have nothing to do with either of them. The structure is similar. We are not even attempting to talk of advertisements and how they have taken over the newspaper. Each item— news or house—has to be taken discretely, and this segmentation and our knowledges of how to perform these segmentations are now forms of sociability. Naturally, if our neighbor needs our help for something, we will do what is needed; we are reasonably altruistic, most of us. Let us remember our somewhat Greimasian rectangular diagram in chapter 2. A fundamental question is how this common altruism turns into its opposite in a flash.

Remarkably, this urban, spatially organized segmentation is imitated by the organized segmentation of the items in a newspaper and the organized segmentation of items in a mall. In a mall one goes from one section to another, one after the other traversing the market itself. In a traditional small shop, it is the shopkeeper or his or her assistant who moves from commodity to commodity, collecting relevant items, and the customer stands at the counter. The commodity comes to the customer. In a mall, traversing this organized segmented space of the market as such, the customer goes to the commodities, collects them, and exchanges them for money at an equally organized segmented space called a payment counter or checkout. The mall renders the market space traversable, and because we traverse that space, we think we are choosing. The phenomenal experience (a tautology) of traversing the space is fundamental to this organized segmentation, which creates a strange sense of order. The first time you go, it is all confusing;

you have to read the signs and find what you want; by the third or the tenth time (depending on retention and protention, and memory), you just walk to the commodity, pick it up, and go to the payment counter, impatient with people who are buying supplies for a whole month or people who fumble about looking for wallets and change and babies and kids running about and are, like the car ahead of your own on the road, blocking your projected timeline and spatial trajectory.

We are comparing here these commodity related matters with what might be called caste segmentation. We have already indicated that Phule and Ambedkar managed, in their own ways, to rearticulate caste segmentation (supposedly four-layered). Phule instituted a *dichotomous* way of thinking about Hindu "society": There were the brahmans, and there were the shudra-ati-shudras (perhaps this dichotomy could be called a "race war," following Foucault of *Society Must Be Defended*).[6] Phule's thought seems to be oriented almost entirely to a projectable present. Ambedkar, however, seems to think of the long-term future, especially in his capacity as a designer of law and constitution (the right to change the constitution is constitutionally given to the representatives of the people). This long-term future is, in fact, imagined in universalist, eternalist terms (law would not make sense as a temporary, contingent, variable historical solution). The noncontingent intention of law-making (it claims to be able to calculate all [but not every] contingency) makes a travesty of time: That is its contradictory and forceful foundation.

It is then Phule who turns out to be more canny, for he does not seem to make universalist or eternalist claims. He is a phenomenal (in both the senses of that word) thinker who is merely trying solve problems that he experiences. Ambedkar, however, makes laws as well (we have already here and elsewhere marked the complexity of that practical gesture). Assuming that the foundation of law is contradictory, it is worth investigating what it might have meant to Ambedkar to be one of the most important members of the drafting committee for the constitution of India. In a contradictory reinstitutionalization of "backward classes," "lines of mobility" were opened up that made the earlier hierarchies traversable. It is the assertion of the traversability of segmented sociabilities that in fact *is* the process of reinstitutionalization.

It should be clear by now that our metaphor of segmented sociabilities and their customary forms is different from the metaphor of hierarchies and society, and similarly, the metaphor of traversability is different from that of mobility and mobilization. The difference consists in assuming, or not, that there is "society."

Where in "society" we obtain "hierarchical social relations," in segmented sociabilities we obtain a relation of nonrelation: We can trade, exchange, have sex with, "the lower," and yet not relate to them in any manner, and that too without using "them" as means to some end (that still is a relation). This relation of nonrelation is similar to the "benumbing" of the sense of touch in a crowded bus or train or any crowded space where bodies touch other bodies involuntarily. That too, we realize now, was a segmentation of the spatial distribution of bodies; since there could be no segmentation, the sense of touch was shut down. Buying something in a mall is similar too— we do not have to relate to the seller except at the time of paying, and the primary producer and seller is usually absent in any case.

We can arrive at this point from another angle. Baburao Bagul's stories have, repeatedly, represented this lack of "society" and the lack of "sociability." One finds only characters "struggling to survive," in fact, characters whose continued living or continued psychophysical sustenance is constantly threatened by other characters. These do not seem to have the privilege of belonging to a society. This reveals, quite powerfully, the club-, clan-, group-, community-like nature of "society": Only some people have the privilege of being members, and this seems equally valid of Western and/or "more advanced" "societies" as well. The man or woman who lives on the street, out of a trolley in New York or Berlin or even small urban places like Tübingen in Germany, does not have the privilege of being a member of "our" society. It turns out that samaj was, after all, a correct translation of "society." Perhaps it is time that we started using samaj for "society" instead in all contexts, to bring out the privileged nature of it all.

My use of the term "sociability" is meant to indicate a phenomenon that is at a level lower than that of "society," which is a much more formalized phenomenon, with several regular features (customs that are "followed" by many people and can even be seen as institutions). My argument is that un/touchability, fingerprints, and sociability form one group, and commensality-connubiality, signatures, and society form a higher-level group. Therefore, it is possible for the former to *de-stitute* the latter; however, on second thought, *de(con)stitute* might be a better usage. Whether the former group can further be de(con)stituted is a difficult question for us, and we cannot answer it, as of now. What is clear at this moment, however, is that since "society" can be de(con)stituted into segmented sociabilities; large amounts of sociologically produced "knowledge" dissipates into unorganized assumptions, dispersed empirical observations, uncoordinated statistical data, upper-caste/class platitudes masquerading as knowledge produced within disciplinary formations. It seems to us that this line of thought can further prove that the totality called "society" never existed except as an unexamined assumption, because these segmented sociabilities cannot be totalized into "society," while keeping the lines of segmentation intact and mostly visible: Either we ignore these lines of segmentation or pretend they do not exist and believe that there is something called society, as if the word (and the phenomenon) was not pluralizable, or we accept that there are many societies (within what is recognized as a society). A society, or societies, and the pluralized form would seem self-defeating.

We can pass through this seeming impasse/ aporia by thinking of customary/ habitual forms of sociability as the conditions of possibility and impossibility for "a society," never ever forgetting that one has to choose between forms of sociability customarily available every and each time we encounter people we do not know—in sum practically inventing a form sometimes customary, sometimes new.

Love or hate at first sight could very well serve as examples of such an invention of a new form of sociability. It should not be forgotten that such daily invention of sociability is a recent phenomenon. In earlier times, it would be far more infrequent since one would meet a stranger far more infrequently. The anonymity with which one must live in urban spaces has this tremendous creative possibility in it: the invention of sociability in everyday life. To appreciate this one must think of a small village in which everybody has always known everybody else. There is no need to invent a new form of sociability since forms of sociability have already been established and have become familiar to us at a very early age. There cannot but be several hierarchies here—men and women, adults and children, men and men, women and women, humans and animals, the rich and the poor, and so on. Education has made just a tiny dent, and youngsters who leave the village for the city usually do not return. Touching a stranger is impossible. There is no need to invent anything.

SEGMENTATION AND TRAVERSABILITY

Segmented sociabilities—by which we mean mainly caste segmentation—do not add up to a whole, and therefore the question of some organizing principle that will encompass and comprehend them does not arise (this is not to state that there is no organizing principle that we can identify, but that such a principle would only have to be an empty function that this or that value can occupy). What remains then, is the question of how to traverse the planes and vertices of these segmented spaces.[7] It has been clear for some time that it is not merely a question of mobility (understood as vertical or horizontal, and mainly in economic terms), but also a simple traversability, crossability of these segmented spaces. It is well known that upward mobility and access to resources have not changed segmented sociabilities, because the section—the cut—that segments sociabilities determines such traversability and crossability as a trespass. These sections/cuts are, again, customary and acquired and learned.

It seems to us also that traversability is different from resistance to institutions. The latter remains preoccupied with/ by institutions.[8] Neither is it a question of leaving institutions—institutionality as such—altogether (it cannot be done in the first place). The reinstitutionalization of dalits at the hands of Phule and Ambedkar and their more thoughtful followers was an interesting step (although they might have understood their own thought and action as political resistance), in the sense that in this reinstitutionalization the earlier institutions had to change. However, because they remained largely institutional, these changes did not reach down to the level of segmented sociabilities. The higher-level, institutionalizable forms of sociability moved ahead, but the actual (actantial?) sociable content lagged behind—they did not render the segmentations traversable. Alterity continued to manifest itself in stereotyped "others" made into empirical beings. (As already pointed out earlier, in chapter 5, the pathological and sentimentalized notion of "identity" came into service of maintaining these segmentations.) When these beings are made empirical, our relation with "others" can easily become psychopathological, and these psychopathologies can begin to guide our politics. This process leaves no choice to either parties but to perform their identities, even to perform in the name of their identities—one can do nothing more with identities than continually assert, confirm, and conform to them. Such identities are not conducive to changes in sociabilities.

Sociability, as a future possibility of being social (without yet having to integrate into a single society), must be thought of as planes (we are making a grammatical mistake here—we are treating a singular as if it were pluralizable—but it is also a figure called hyperbaton) that are segmented/ connected by vertices that are themselves traversable. Inasmuch as all the segmented socialities occupy the same world, they have to be traversable. All the segmented planes occupy the same (not merely similar) space-time.

The alterity that the vertices—sections/ cuts—introduce needs to be taken into account. That is why we have insisted that alterity is a fundamental element in sociability (whereas the notion of society tends to diminish, if not minimize, alterity). It should now be possible to state that caste segmentation is an issue of alterity. There are several societies; we choose to belong to this or that, but sociability is a matter of how we relate to those we do not know. We have already argued, from the beginning almost, that touching others introduces us to alterity in its materiality, in its palpable phenomenality.

These segmented sociabilities become traversable if our relation to alterity is open with a welcoming hospitality. This necessarily involves that we unlearn our privilege *and* our disprivilege as well as their reversibility (we shall no longer be able to authentically assert our "marginalized identities"). The "globalized" sociability that capitalistic expansion promises can now clearly be seen as this one, and one alone, caricature of sociability, since "globalization," while constantly invoking otherness (of something called culture, sometimes called cultural difference), does not offer any description or analysis of alterity. We have already indicated this earlier and need to expand on it further.

The traversability we are attempting to figure depends fundamentally on the notion that all of us occupy the same space-time. Our understanding of temporality might differ; our understanding of space might differ, but it is the same, the one and only space-time that we have and inhabit. "Society," the existence of which we have denied earlier, can now be construed (not constructed) as a futuristic ideal that is realizable precisely because it is rendered difficult by the segmentation of sociabilities. It seems to us that although these segmented sociabilities are "different" from one another, the vertices that segment/ connect them are identical: the vertex or vertices, of alterity.

We cannot enter here into the very difficult philosophical problem of whether alterity itself is the same everywhere in the space-time that we inhabit. We could suggest that it is possible that alterity too is divisible, internally differentiated, traceable. Is the vertex that segments/ connects the brahman of this or that branch from/ to another brahman of that or this branch the same as the vertex that segments/ connects the brahman from/ to, say, a *kunbi*? Or a *mahar*? Are the vertices that segment/ connect the various "subcastes" the same as this one? Although we are not solving the philosophical problem that these questions pose, it should be observable that sociabilities unite in the face of what they perceive as the "more other" than the "less other": Thus, brahmans of all subcastes unite in the face of the "lower caste" and so on, and Hindus unite against "Mussalmans" and so on. And all of us might unite against whoever is the currently nationally designated national "enemy." Thus, it becomes necessary to admit the notion of the "totally other," absolute alterity. That we are not able to know or even guess anything about absolute alterity connects it to futurity. It is sociabilities, when they traverse the vertices, that make possible the "social." Society is still a little further away, in the future, and socialism/ democracy further away still. Thus, we get what seem to be four layers: sociabilities, sociality, society, socialism/ democracy. This is somewhat fanciful but makes us aware how much of a higher-level phenomenon "society" is—something that we have always taken to be the lowest fundamental level. From this perspective then, the possibility of being social is premised on what we have called the traversability of segmented sociabilities.

Such traversal—even if it is across only one vertex—opens up the realm of "freedom" in the sense that it enables us to cognize other sociabilities. We cease to be tied down only to those sociable gestures that we are able to invent or use when confronted with others. We begin to cognize the possibility of using other sociabilities, freeing us from our habitual invention of our own kind of sociability (and above all, freeing us from our own "identity" [I say *namaste* instead of *zohar*]). In this process one also comes to learn that one is an other for some other person, and the other person comes to learn of our own alterity for him or her. Alterity thus ceases to be an opaque wall of division or incomprehension but becomes a hinged turnstile or door which can open both toward the inside and toward the outside. (The movements of temple-entry and the movement for access to water at Mahad [both led mainly by Ambedkar] we can now understand not merely as "protests" against "upper-caste" domination or exploitation or even hegemony but as attempts at articulating the possibility of traversing to other sociabilities.) These movements not only brought the "backward classes" into spaces where they had not been ever, but at the same time made the "forward classes" cognize their own sociability as something that "others" could accept.

The doors that the "forward classes" thought destiny had made such that they could only be opened outward[9] and thus could not be "forced open" were now pushed open through the persistent demand that the traversability of these segmented sociabilities be acknowledged and admitted. One door opened outward (the "backward classes" came out of their own segmented sociability), and as a result of that door opening, a door was pushed inward and opened.

Yet again the notion of segmented sociabilities (in their traversability) allows us to question the notion of society. There is another way of arriving at this question: If in most socioanthropological and historical accounts, society is irreconcilably divided by caste, class, race, gender, if it is thus fragmented, by what logic do we need the notion of a one and a whole society? As is evident, the notion of something divided, something fragmented, assumes the idea of a whole (usually located in some rememorated "past"). How do we recognize a fragment as a fragment unless we have already experienced or imagined an "idea" of a whole? (Is the idea or experience of a "fragment," then, always already idealistic?)

A good (performable without losing one's ethic) experience of "society" is precisely what is missing, precisely in these divisions/ segmentations/ segregations of caste, class, race, gender—not to speak of the incredible amount of everyday violence and singular but common massacres and group killings that mark our past and present. It seems logically necessary that the notion of society be discarded, since it seems to be no more than a dream of a wholeness that is held together only by some magic of unexamined belief and assumption. We know how quickly that dream turns into Khmer Rouge and Pol Pot, into Nazism and Hitler, into Biafra and Rwanda, into "ethnic cleansing," Bosnia-Herzegovina, Nagorno-Karabakh, into Gujarat Riots, and more currently, the Rohingya genocide—into wars all over the globe—into the sovereign power to kill—one person killing another.

This attempted erasing of alterity—by destroying its cognizable manifestations—is almost exactly the opposite of the freedom that traversability promises. We have already seen, in the riots of 1992 in Mumbai and Surat and other places, how a neighbor suddenly turns into "the other" and thus into "the enemy" and thus turns into a killable animal whom you must kill, and have killed already anyway, since you thought you had to kill me because I am your enemy. The other is polytropic, and by that logic, also polymorphic; therefore, this particular form of alterity, namely, me, is the point of application of the power-to-kill.[10] But I am nothing more than a particular/ specific/ perhaps even a singular manifestation of alterity.

ALTERITY, TRAVERSABILITY, AND LANGUAGE

There is a knowable alterity and an unknowable; there is a knowable traversability and an unknowable. We acknowledge knowable alterity when we say we understand and know differences in cultures, languages, modes, mores, and so on (this does not necessarily make us less violent [please note that we are not using the prefix "non"]). We acknowledge knowable traversability when we talk of mobility, mobilization, migrancy, diaspora, and so on (these conditions do not necessarily make us happy). Unknowable alterity we call the totally other. Following the same logic, unknowable traversability will have to be called the totally traversable, and we suddenly discover a positivity. We have just called it into being, in saying it.

Total traversability is a forever present possibility. The vertex that seems totally untraversable is that of death. Even that—the vertex that segments existence and extinction—is traversable in language and memory and imagination. We often speak of the dead as if they were alive; we remember them; we imagine what they would do had they been alive. We need not consult here anthropological material on how the death of X is managed by the living samaj in this or that primitive tribe/ samaj/ society or "institutionalized religion," or "advanced" culture. Although such material is instructive, it is not of much use to us here.

It is necessary to thematize total traversability specifically as "freedom," for it becomes clear now, almost at the end of our argument, that without the notion of freedom, "social" change (the second element of the series, sociability, social, society, socialism/ democracy) is not possible. Configuring freedom as total traversability has the advantage that, inasmuch as traversability is movement "from" a point "to" another point, it encompasses both "freedom from" and "freedom to." Anti-caste struggles have, until now, mostly emphasized "freedom from"; it is necessary therefore to focus on "freedom to." We already have described traversability as willing hospitable sharing of other unfamiliar sociabilities. Configuring freedom as traversability also allows us to avoid the mistake of thinking of freedom as if it were some substance or attribute of substance that X or Y possesses/has, or some continuously present structure of inherence, or, for that matter, some invariable concomitance of a being and a freedom. We understand freedom as possibility of a real, phenomenal movement in space/time: traversability.

As beings/ entities that we always are, our fundamental movement is toward death, a point in space-time when we cannot move by ourselves. We can still be a "thing" and be "rolled around the earth's diurnal course" (if we are buried a few feet below the earth [at the center of the earth, there would be no diurnality]).[11] If we are burned after our movement ceases, we will be somewhere, but not identifiable in space-time as a place where we happened (happen is related to chance, randomness), where we were (where our life took place). The body as the place where we used to happen is turned to ashes, and it cannot re-congeal into a place where we could happen again perhaps. "My body is so / scattered that I cannot go to it."[12] By the same logic, in case you think we have forgotten the heading of this section, there is a knowable language, and there is an unknowable. It is a strange thing, is it not, that we can talk about something, some meta-thematical thing X as unknowable in language, but we cannot speak of an unknowable language? We have already indicated that "language," whatever that means, whichever it is, permits us to speak of nonexistent things. Dead people live on in language, like living people live on with language.

It is also a strange thing that whatever language/s we know, that does not matter for our bodies. Languages do not add up, in case we know them. Although it is tempting, we cannot enter here into a discussion of evolution (body, genes, ageing, dying, survival), since that notion of evolution was initiated by language studies of the eighteenth-century European kind, where they could not explain resemblance without pending from familial metaphors.

Let us be elliptical rather than cryptic: If caste is predicated on birth, the de(con)stitution of caste might need to be predicated on death. This seems to be a logical requirement. However, since every beginning of life is also the beginning of death (the moment I am born I begin to die), living and dying cease to form productive oppositions. Thus, it would be necessary, to predicate caste on death as well, since birth and death are simultaneous processes, not only simultaneous, but also processes that take place in the same place: our bodies. Does this help us in the de(con)stitution of caste?

It does not, because we have already observed, taking a hint from Phule, that caste seems to be nonbiodegradable. Therefore, our biodegradability might not be methodologically useful for our purposes. Where can we, then, begin such de(con)stitution? Where do we find our point of departure? It always has been simple; the point of departure would have to be the point where we are: We will have to move away; we will have to traverse some space—intellectually and metaphorically. We will have leave point X (where we are), for some other point that might even not be on the same plane and might be difficult to traverse, because we might not be able to identify the vertex that segments us from it.

Let me count and list what we now dispossess (we stand alone in a space of transition from our world to some other, mostly seen as inimical, like Shevek the physicist in Ursula Le Guin's *The Dispossessed*): We have dispossessed ourselves of caste as the most significant marker of injustice in "India." We have dispossessed ourselves of the varna "system." We have dispossessed ourselves of "society" and by the same token most "sociology" since it does not seem to be able to study anything more than "society." We have dispossessed ourselves of (the choice of choosing) customary forms of sociability. We have dispossessed ourselves of institutions. We have dispossessed ourselves of "Hindu" metaphysics articulated in Sanskrit (which we do not know anyway; we never possessed it; nevertheless, we have given it up). We have dispossessed ourselves of the Constitution, since we know that it usually is reduced to Law and to the Penal Code.

Lest it be thought that in this dispossession of the Constitution we are disregarding Ambedkar, we would like to say that his faith in it, and in law, has been betrayed, and in hindsight, perhaps even misplaced. In any case, he has been, and is being given so much credit for so many things that he would never ever, even after dying, have any debits, he cannot now, and would not have been able to when he was alive, spend all those deposits in his account.[13] For us to be able to attempt a de(con)stitution of caste, we will have to move from where we are. We will have to traverse somewhere else, some as yet unknown, but existing (present indicative) space-time.

Let us traverse a few thousand (about four, we think) kilometers, and go to Japan, Hokkaido island, in particular, and to Fukushima (quite far from Sapporo or other urban centers on Hokkaido). It seems that there are, and we are using well-known and studied phenomena, people called Ainu in Japan, who are discriminated (against). (The Fukushima catastrophe had an afterlife and care, and who do you think was employed to go in first without full protective gear, which the TEPCO employees were given? The *burakumin*.)[14]

Let us go west as well: Let us now go to Yemen and come to know that there are the *al-akhdam* with "negrito" features on their bodies. We know, with even a single reading of a newspaper, that there are many such violations of cultured bodies. We have dispossessed ourselves of the customs of commensality (we wrote commensurability but changed it) and connubiality a while back. We might now have maneuvered ourselves into some position from which we can move to somewhere else, some future.

8

RECAPITULATION WITH VARIATIONS

PROMISCUITY

Such a promiscuous word *caste* is.[1] It is used in so many ways and brings so many references, and it delivers, after some yielding and coaxing, so many meanings. It does not remain faithful/pure/chaste to itself ever; it is clear that no caste is celibate.

This is true of most words, but it is more evident now in this one. The same is true of the word *jaat* in Marathi. It is all mixed up beforehand. It could even be said that it is not one word but many.

Also, such diverse and heterogeneous practices as eating certain foods, marrying certain people, sitting in a certain way, dressing in a certain way, speaking in a certain way, addressing others in a certain way, performing certain rituals, voting in elections, emplotting histories, killing people (and this is not a *gradatio*), and so on, are done in the name of caste that we might have to admit that it is not one thing but many. There is a complementary and self-reinforcing movement here: Things are done "in the name of caste," and things are also explained, post hoc, by the name of caste. The sum total of these things is: "caste." It does not seem too much to say that caste enables us to unite and totalize (make whole) rather diverse and often unwholesome things. It begins to operate a little like a proper noun: The various things that I am and do and say are called by my proper name, uniting and totalizing them under one name; similarly, various things that we do are given the name of "caste."

So the promiscuity—the beforehand-mixedness—of "caste" gets obscured. It is important to remember here that Aristotle in his *Poetics* points out that unity of character does not necessarily mean unity of plot. Thus, on the one hand, nobody seems to deny that there was mixture of blood (through reproduction) from the so-called Aryan invasion/migration on; yet, on the other hand, there is a rabid insistence on maintaining purity of blood in reproduction. Derrida's observation, in "Structure, Sign, and Play," that "coherence in contradiction marks the force of a desire," is apposite here. It is necessary to identify and analyze what the desire is, here, in this coherent contradiction, as well as the mechanism that functions to maintain and work it.

For there to be purity of blood and birth, it would be necessary to reproduce only with someone related to you by that metaphor of blood relations close and distant (we are told by scholars that in ancient Egypt the pharaoh and his sister could have sex only with each other so that "pure" babies were born and could become pharaohs and their sisters).

This would mean that one could not allow the incest prohibition to spread/extend beyond a certain limit. The incest taboo, taken usually as a constructive restriction in sociology and anthropology, itself had to be restricted, inhibited, circumscribed, for endogamy to be possible and thus actual (thus it was centripetal—it could not be allowed to be centrifugal; it could not be allowed to run away from the center of the incest prohibition).

Endogamy has to negate an absolute incest prohibition—no sexual reproduction/relations between those related by blood in any manner (space-time whatsoever). It would then follow that endogamy and exogamy are not antonyms, certainly not opposites, and definitively not binaries: Endogamy negotiates between the absolutization of the incest prohibition and exogamy. We might want to think about the words *exogamy* and *promiscuity*. Incest is institutional; promiscuity is random.

One must remember here that the so-called incest prohibition is tenuous: Sexual abuse mostly happens within the various degrees of incest. The incest prohibition works as a prohibition only for "socially legitimate" sexual relations. Those who think that the incest prohibition is the foundation of some form of society or culture, or regulation of sexuality or sexual relations, must make themselves blind to the fact of sexual abuse within the family in order to maintain their belief; at the least they must treat sexual abuse within the family as an exception—rendering it negligible. Incest prohibition then seems to be mixed with, contaminated by, incestuous sexual abuse. It is clear that the incest prohibition itself has a temporality to it—for if we go back in time sufficiently, all of us have common ancestors. This is the other way to contain it—in finding common lineages we do not go back sufficiently; otherwise, all marriages and sexual and reproductive relations will have to be deemed incestuous. This is how we protect our rather austere and ascetic promiscuity, using our genealogies only within a certain time frame, not going sufficiently backward even though the genealogies themselves might: We work hard not knowing the Mitochondrial Eve, the mother we all seem to share metaphorically and genetically.

LOGIC AND RHETORIC

We have emphasized enough already that there are operations of metaphor and metonymy in the functioning of caste-as-system and caste-as-practice. It is perhaps time to see how exactly these figures can be made to work as guidelines for conduct, metamorphosed into systems, symptomatologies, symbols, explanations, heuristic devices, cognitions. Even more than metaphor, it is metonymy that calls for attention. In chapter 5 we mentioned Ambedkar's notion of caste as a "unit of guilt" (one person violates/ transgresses a norm and the "whole" caste is punished). By the same token, it is also a unit of justice: But should not these depend on logic and not on some notion of figure, *alamkara*,[2] some tropology?

What allows a figure to function as a logical principle for conduct? It seems that the one place where logic and tropology fleetingly, almost tangentially, meet each other is in the cause-effect metonymy, which is often taken as a valid logical inference.

The issue gets further complicated when we realize that a lot of assertions about caste are not, strictly speaking, syllogisms but enthymemes, forms of persuasive argument that depend on shared opinions (and not logical inferences), for example, "Among the Hindus, only lower-caste people eat beef," or symptomatologies that produce commutable/ commutative sentences: "She eats beef because she is lower-caste" commutated as "she is lower-caste because she eats beef." An "inference" is drawn, but it is in the service of persuasion, and not of "proof" or "demonstration." Yet it behaves so like a proof/ demonstration that it looks like an irrefutable logical inference. It seems that this is really what gives power and efficacy to all the caste prejudices that operate.

We suggest that metonymies gather their power because there are contiguities in space-time, and this space-time is the same that all of us inhabit, and it is traversable (contiguous, unless we have been rendered immobile mentally or physically). In a certain sense, metonymies seem to depend on a certain notion of apophantic "sameness":[3] We do not want to touch certain bodies because the hands that are attached to these bodies are the "same" that transport excrement from our toilet to some other necessarily unknown place; they are the "same" hands that process animal fat, touch and eat dead cattle, and so on. Such a notion of "sameness" must ignore the fact that these hands perform other "functions" as well (that is where the apophansis should fail, logically), used for other purposes as well: These hands are the same hands that must be caressing a lover or a child "even as we speak." How did you separate the hands from those bodies to make them into metonymy? Because you saw the hands attached to bodies that take away your excrement.

(Other people fuck; we make love, no?)

The ignoring/ignorant apophantic signification of the "sameness" across diverse actions seems to congeal into an identity that is transcendent to these diverse actions. This is what makes it possible to attribute a certain "caste-identity" to persons. This transcendent identity then begins to function somewhat like a transcendental signified, producing and regulating and controlling a performative semiology and semantics of "caste." A word that is but one word among all the others becomes something like the center that controls the play of other words.

IDENTITY POLITICS

Our thoughts on caste are caught in an aporia, possibly of the variety that Derrida pointed out in the context of the concept of the "sign," in *Of Grammatology*: How can we criticize "caste" without giving "caste" some semantic operationality/functionality in our criticism?

Naturally, we might not be able to pass through this aporia by following the protocols that Derrida follows in his context (we are not able actually to follow them). It would be interesting though, if we could begin to write ~~caste~~ and follow those protocols.

What we could do is to attempt modestly to see clearly the aporia itself: If the anti-caste struggle is a struggle to end "caste," then any evaluations of "caste-identity" would have to be taken as detrimental to the end of caste (since we are trying to imagine a time and space in which and where "caste" has no meaning or value whatsoever). Caste would have to have been thought of as irrelevant or insignificant each time, by everyone and anyone.

However, increasingly, in dalit literature and dalit discourse and every-day politics in general, one encounters positive assertions of "caste-identity." When coupled with our individual narcissism, especially in upwardly mobile dalit "organic" (organon, *yantra*,[4] instrumental, but often understood as a natural-part-like, therefore more authentic than anybody else) intellectuals, this makes for "political" situations that only can be schismogenic.[5] (Ambedkar's metaphor, in his 1916 essay, of closing doors seems to be a pre-monition of Bateson's notion of "schismogenesis"). It would not do to find alibis in the behavior and conduct of brahmanical people, and their own assertions of their own "caste" (purity). Neither would it do to celebrate one's own caste as something positive.

Over and above this, there is the confusion between "politics" and one's own upwardly mobile ambition and various obstacles to its fulfillment (of-ten enough, oppression/ domination is confused with such obstacles). This allows us to treat everything personal as political, as if only the personal is political (this is what permits us to be political in every encounter with someone of the "upper-castes," and yet no social change takes place).

Over and above all this, personal and impersonal collective emotion is narcissistically planted onto "caste" (pushing the center out of the circle), easily rendering others as enemies.

Since this aporia cannot be avoided, we might want to consider the pos-sibility of entering the aporia in such a way that it rejects us by its own logic and sends us away on a different tangent altogether (traversal). It seems that one way of doing this is to give up the notion and practice of "caste-identity."

We can make some argument here. Inasmuch as "caste-identity" is attrib-uted to us by others, and inasmuch as "caste" is to be ended, we would have to give up not only the "caste-identity" attributed to us by others but also the "caste-identity" that we assume for ourselves (attribute to ourselves because of our personal pasts and attributes [narcissism]). Inasmuch as we need to unlearn our privilege (because it is pre-learned), we also need to unlearn our un-privilege (because it is pre-learned). Such unlearning needs be in the educational mode and not in the political mode. This is an important issue, since if the unlearning is to be (taken to be) held before and (apprehended) in the political mode, we often mistake it for upward mobility. This happens because politics themselves are, in such cases, understood as economical politics and not political economy, which is rarely practiced now.

They have managed to mathematize economy, allowing us to mistake (substitute) upward mobility for "social" progress—we make more money, it seems, and we are happy with that, thinking we have managed to go beyond caste.

DESTITUTION

We used the word *destitution* earlier and later modified it, backtracked, got scared, and suggested that perhaps *de(con)stitution* might be a more adequate word. We were wrong: The destitute do not, and cannot, make any statutes; therefore, the (con) has to be given up. Neither can the destitute undertake any "destruction" of any metaphysics (Heidegger); therefore, no nice equivalences can be set up between destruction : de(con)struction :: destitution : de(con)stitution.

We are witnessing a systemic double bifurcation in a system: There is a bifurcation[6] from de(con)struction, and from subalternity, though the difference between subaltern and destitute is still obscure to us. We need some fiction—one that is not heuristic—to get us out of this.

There are, of course, no lotuses, no elephants, no palms, no clocks, no cobras in this story.[7]

There is almost nothing in this story except us: you and you and you. That's us. It is rational to be scared when we don't know where we are going, meaning you will understand what we are saying, so you won't be scared, but we are.

Then there were you and us, suddenly. Let us, carefully, enter the fiction. This is crucial. And let the fiction of our togetherness remain unspoken between us, because we are to be many, not just us, you and me.

We hear ourselves only when speaking to others mostly: When we talk aloud to ourselves, it's close to madness, every time. We have been talking too much to ourselves, sharing our suffering, saying to ourselves that they cannot understand. The us and they have been variable, but still aporetic.

The melody must destitute itself musically.

The invitation of the us who knows (us as a singular grammatical number)

The single note after the melody has destituted itself, the note of the us is all that remains:

"I'm Nobody! Who are you?
Are you–Nobody–too?
Then there's a pair of us!
Don't tell! they'd advertise–you know!"[8]

What can we say then, on this destitute (infirm) ground? We can say very little, do even less, it would seem. We could ululate: us, us, us, all of us, us. Not us and them, not you and me making up a divisible us. Just us.

We have been told that we must constitute us, ourselves, in a limited manner. Me, my family, my friends, my clan, my caste. This limits the us to ourselves merely. We must reject this limited us.

We are not inviting you and ourselves to some vague "us human beings." We need to stretch the us to its limits, and we might find that the real limits are still further away. That is the impossible limitlessness of us that we need.

It is clear that institutions imagine limitlessness in a mathematical manner, but we who are destitute can only imagine limitlessness not as something to be conquered, but as something to be explored, understood and incorporated every time: We're nobody who are they? Are they nobody too? Then there's much more of us. Let's tell, we must tell others they are us, you know.

REARTICULATION

We should not attempt that which is in principle (de jure) and in practice (de facto) not possible: To imagine the future as some state that we are going to reach by taking certain steps in the present, especially if we are attempting to imagine a future that is casteless, classless, and so on (we have not used the word *society* here). The logic is very simple and of humility: if we could imagine/specify/ know the steps to be taken (actions to be performed individually and collectively at the same time: "politics") that would take us to the desired future, surely one of us would have already taken those steps. The us requires that. We would be there already. This is not to deny that actions that one or many of us take change our future, but to ask how we know the future will be changed. . . .

Future, here, is deeply encoded by memory. I remember seeing a film, reading a book, meeting a person, listening to a talk that changed my life.

Many things seem to have changed our lives. This is strictly not a question of numbers, for the us is not an arithmetical number at all. This is also an illogical expectation that because our life changed because of something, we will be the cause of changing something. There are no such repetitions and/or reenactments.

The rearticulation at the level of economic and social destitution does not seem to have changed things as much as expected. So perhaps a rearticulation at the intellectual level might be attempted.

De(con)stitution, still, might be a useful word if treated as a pun (paronomasia, admonitio, related to polyptoton), ignoring the etymological bag in which we put lots of our material, rather than as a concept/ metaphor: We will have to give up our constitutionalism to reach some space-time from which we might be able to begin to give up "caste." Which would also mean that we will have to stop looking up to the state to protect minorities; this is, at the current juncture, dangerous.

Those of us who are proud of the constitution and uphold it because Ambedkar wrote it will have to rethink their relation with it and with him. We should have said this earlier, the moment we used the word *de(con) stitution*.

In such work, atheism will be indispensable. Without atheism, nothing much will happen; nothing will change until the us gives up theism. The us is now at that state of de(con)stitution. The us is now left only with sociability. The obvious question is: What if "others" refuse to be sociable? Refuse to be convinced by rational argument? Insist that the way "they" have been doing things is the correct one?

The us is then left with a very bleak view of the world, far bleaker than most dystopias have offered. Yet the us remembers that others believe in something called society and that the space-time indicated by that word has been expanding: partly as modernity, partly as the anti-caste struggle's success, partly something else that we do not know as of now. That the space-time of "society" has been mostly urban—that has to be admitted.

That that space-time is mostly available in "social" groups that have middle- and upper-class markings and behavioral habits—that too has to be admitted. That that space-time is rather constricted, in terms of numbers—that too has to be admitted. That the existence of that space-time does not promise much in terms of change—that too has to be admitted.

Over and above all these admissions, that openness that we have called sociability, and that which sociobiology and ethology call "altruism," does exist. Like most things that exist, it too can be reproduced, trained to be better at what it does and thinks, and made to reproduce itself later in a better version. Birth is, in that expanded sense of time (evolutionary time), one iteration in a recursive process, or unknown algorithm, and not the final determinant of anything at all, just as death is not. "They" will die too, perhaps after they kill us. Inasmuch as we are living and dying together at the same time, we are also cause and effect together at the same time. But alas, there, although space is traversable, time is not. Can the us change the direction of time? Let us hope.

PRACTICAL DEONTOLOGIC

"Nothing of this is going to work in real life." No it is not about real life there isn't a thing called real life it's all in your mind dear nonsolipsistic reader because you have accepted the fact of your birth and not accepted the fact of your death you are only an interval a diastema the christians meaning paul and agamben say it seems and your experience of your own existence is so limited you do not even know yourself and you claim to know about real life and I ask you how many people do you know maybe a hundred and you make claims about humanity and the lot including animals where as some of us are merely trying to become human from being an animal and some of us trying to become good animals from being horrid ab-hominable humans

No as FaNOn said as we have said NOw

That is a repeatable syllable

We are many, no?

That needs repeatability iteration patience responsibility critical thinking and these are not platitudes we are making because we have nowhere to go now it might seem but actions We did it, do it again and we have to do it ourselves without any clarion calls or programs for action from anybody or anywhere

that we are here alive and talking is a good point to begin to change

When we begin to change those that we call others who are a part of us will have to change

agency we take for a change that is loose and slow and not big denomi-
nation MONEY.

Just as we had to because they who are part of us did

Age old known wisdom that if (deontic logic, let us make it funny and
call it deontological)

we want change we must change this is thefigurecalled polyptoton it
seemsoris at least its readable

as the figure polyptoton since we are educated it might seem

The us has an ancient voice the self-same sound that you should now here
and

The oracle and the ghost said what was needed to be heard aloud because
it was known but

not acknowledged

the amount and kind of known everyday and exceptional violence that
is present in the

world tells us that there is no society because it only seems to be a privilege to belong to society

perhaps there isnt even sociability since we cannot even think about
violence that is not known to us

andtherulesthatyouhavelivedbywillbecomecleartoyouasmerer-
uleswheniforgettopressthespacebarand

nowyouknowhowtosegmentwordsbutyoucannotdoanythingexceptkillkill
killandkillagainbecauseyoucannotacceptyournarcissismandbecuse thats all
you know how to and so on, no? That there are otherpeopleintheworldallyou
wanttodoisreplicateyourselfthereforeweneedaparalleltodeonticlogicthat

hasbeena change an option

andyoumaynotrealizebutthissentenceisendfocused

becauseweneedtothinkaboutwherewearegoingwhatiscaste'send

weneedtokillthisgodforsureifwearetogoanywhere

maybedistrictnineandthealienshipanywhere

is

better

than

here

iamthenobodywhoaretheyou so let us go and do something

like thinking

i like thinking do you i like thinking, think, with you do you like thinking with me because
for you i am the you and i liked thinking with and the thinking of you,
think
thats my animality it includes thinking
with you thinking
it must happen many more times because why stop at two, we are many
there are many that
are us as always no?

WORKABLE FUTURES

This us may not take us anywhere, that is always possible. But we have to
try. It will not do
to leave things alone. We have to go somewhere, don't you think? We have
destituted ourselves in
many ways, perhaps that reveals to us our intellectual destitution: that
we cannot think of where to
go and how. The answers are easy enough, but we have given up that ease.
What? We have made ourselves think and write and read through all this
to arrive at platitudes? No. We have, among all the destitutions, also destituted ourselves of society and the violence it brings with it. We are not seeking "humanity." We seek an animality that kills only for food, not for
abstractions. We seek an animality that lets be, one between "humans" and
those creatures that humans call "animals," and again why stop there, perhaps we need vegetational life: slow. We are many. Our job is not merely to
initiate social change (without being vanguardist activists and leaders), but
perhaps even more importantly to maintain social change through time and
history. To think that we have a goal, which once reached, our job is over!
Social change needs to be permanently maintained and nurtured. We will
have to keep changing; there is no stopping here. We need to think what we
will have become after the "end of caste," what we will say, what we will do.
Will we be good persons? Must it not always come that will we be good, and
not merely will we be happy that does not seem possible since we cannot
make all of us happy in some way. Us. That's all that is left, what are we going
to do with the us, with ourselves.

Let's do the inevitable initialization: Let us forget everything. The annihilation of past is
necessary before the annihilation of caste can happen. It will not do to
cite radical or orthodox
instances from the past in order to initiate social change. The past itself
will have to be forgotten for
we remember far too much of it, we have been made to remember, rememorate too much of it. So
much so that we have been unable to imagine the future. This also helps
in desacralizing time, stops
us from living in an eternal present of a non-Christian interval (*diastema*)
between a mythical past
and a metaphysical/ spiritual future (one which denies our certain and
literal death). We need to
forget that we are "of" this or that caste. If we do, there are so many more
people to embrace and
touch with affection or have sex with, or to reproduce with. In order to
become someone else we
need to forget what we are and have been.
What shall we do then, to assist the *al-akhdam* and the *burakumin* and
the *ainu*? (We will
have to learn Arabic, Korean, and Japanese—the dialects that some of
us use, not the travel guide).
What can we think and make ourselves do for those of us who are much
like us? What can we think
and make ourselves do for "the poor"? What can we think and do for
us? They are us. What can we
do for those in Rwanda who were called "cockroaches" and killed? What
shall we do for destitute
African Americans? (We will have to learn American English as one of
our dialects). For destitute white European women and men? (We will have
to learn German and French and Spanish and Magyar and Czech and Slo-
venian and Serbian and Croatian and Chechnya Russian of some flavor and
Yiddish and Persian, to name a few metonymically).
There always is somewhere to go, and that is better than being here and
now. Let us go away

from ourselves to find ourselves somewhere else in the world, and in the future. Let us do something for those of us whom we call others. Let us also do something for those we think as dominant/ dominated, and those who think of themselves as dominant/dominated.

We are to help us, without god/s. We the People?[9] Shall we constitute the us here, on this

deconstitute ground?

FOUR VARIATIONS ON A THEME BY ANTONIO GRAMSCI

(with these, after this preparation, we take part in current [August/ September 2016–2017] events around caste: The only problem that remains, and conceptual thinking cannot solve, is how to transform these constatives into performatives, propositions into acts of knowledge and belief and politics)

1. *All human beings are dalits, but not all human beings have, in society, the function of dalits.*

(possible performative: *everyone* admits that they are dalits—"we are all dalits.")

2. *All human beings are brahmans, but not all human beings have, in society, the function of*
brahmans.

(possible performative: *everyone* admits that they are brahmans—"we are all brahmans," and ditto for the rest of the variations)

3. *All human beings are women, but not all human beings have, in society, the function of women.*

4. *All human beings are men, but not all human beings have, in society, the function of men.*

The next page is left almost blank for you to write your variations.

The title of the next section is a misquote from Louis MacNeice's "Prayer before Birth"

WE ARE NOT YET BORN

CODA

And always the loud angry crowd,
Very angry and very loud,
Law is We,
And always the soft idiot softly Me.

—W. H. AUDEN, "LAW, LIKE LOVE"

I have written this down as it came to me, off and on, since 1998. There must be in/consistencies in argument, phraseology, and ideas. In lieu of an apology, I would like to say here that writing this has been an exercise in style as much as an exercise in thinking almost by myself, and it has been hard. I have not changed, mostly, what I wrote a long time ago. Changes were made only to make the text readable.

I add this endpiece so as to relieve some of the dryness and impersonality of the style of the main body of this text and to come through as a person. When I say "an exercise in style," I actually mean it, for working out style it is that enabled me to think what I have written down. I am not the person, really, who should have written this, but I could write it because I followed the thread of style to find my way about, to find an argument, perhaps even suggest a few new things.

The original idea was to attempt a Lévi-Strauss-style diagram of relations of touch. After that it kept growing in my mind, in all directions. In order to make sense even to myself, I had to choose only a few directions, and that is where my utmost desire to attempt writing in an ascetic, minimalist style dovetailed nicely with what I thought needed to be said about touch and caste. I am sure attempts at analysis of caste can also be made using some other sense or notion, for example, the visual or, even more promisingly, speech styles. That book, when one of us yet unborn writes it, will be worth more than this. "The tea-leaf in the tea-cup is the herald of a stranger," as a Louis MacNeice poem has it.

Urmila Bhirdikar, my partner, was a strong supporter all through, and I could not have written this without her. She always knows what I need, and gives, which is more than what I, or anybody else, can say about me. We read the draft aloud, and she made several valuable suggestions to change the style and content. Whatever readability and consistency this text has, comes mostly from her suggestions.

Anupama Rao and Arvind Rajagopal have always patiently read what I sent them and encouraged me tacitly, tactfully.

Amlan and Aditi Dasgupta, V. Sanil, and P. Udaya Kumar have always encouraged me with their expectation that I will write something interesting. Rajeev Patke too has always expected me consistently to think and write well. All of them have also been intellectual interlocutors in many ways in other contexts as well, and I am deeply indebted to what I learned from them (mostly without their knowing). Supriya Chaudhuri and Probal Dasgupta and Gopal Guru have always shown an interest in what I was thinking and writing, and they have supported this attempt without really knowing they were.

I would also like to acknowledge the intellectual probing and opening of possibilities I was generously given in some conversations and some correspondence with Gayatri Chakravorty Spivak.

I am also grateful to those who helped me understand some of the material in Sanskrit: Ramakrishna Bhattacharya, V. N. Zha, Shraddha Kumbhojkar, Ujjawala Zha, and Pradeep Gokhale.

Like many others in Maharashtra, I too am indebted to many conversations with the late G. P. Deshpande and the late Ram Bapat.

I thank Thomas Lay, of Fordham University Press, for his consistent and patient long-standing support for this book; Eric Newman, also of Fordham; and Teresa Jesionowski, my copy editor, for her precise suggestions. I also thank Ishita Mahajan, a former student at Shiv Nadar University, for assistance in making the bibliography, and Sreejata Guha for preparing the index.

I am sure I am indebted, in unknown ways, to many other people (including some whom I take to be my enemies), and I acknowledge here that gift-of-debt in my ignorance.

I dedicate this book to all of us whose stories we will not hear or read in print because they were killed in caste-violence and thus could not find any workable future at all. As always, knowingly or unknowingly, in memory of the dead, and in anticipation of the unborn, I hand this over to you, as me, in my name and with my unsigned signature.

August 2017

NOTES

FOREWORD BY ANUPAMA RAO

1. A brief outline of a set of interconnected scholarly positions around caste will illustrate my point. These consists of the argument that "caste" took on a specific form as a modern construct in the colonial period; that an elaborately articulated colonial sociology of knowledge evacuated from caste its relationship to state formation, territoriality, and shifting forms of power; and finally, that colonial infrastructure and ideology effected a profound transformation of caste, which consisted of focalizing caste's social and political centrality while rendering it an aberrant, or insufficiently modern, nonindividuating rubric of social life.

2. Anupama Rao, "Anticaste Thought and Conceptual De-Provincialization: A Genealogy of Ambedkar's *Dalit*," in *Postcolonial Horizons*, ed. Jini Kim Watson and Gary Wilder (New York: Fordham University Press, 2018).

3. Aniket Jaaware, "Destitute Literature," published as the First Annual Jotirao Phule Oration (University of Mumbai, March 2012), 33.

INTRODUCTION

1. arrogance (n.) c. 1300, from Old French *arrogance* (12 c.), from Latin *arrogantia*, from *arrogantem* (nominative *arrogans*) "assuming, overbearing, insolent," present participle of *arrogare* "to claim for oneself, assume," from *ad-* "to" (see *ad-*) + *rogare* "ask, propose" (www.etymonline.com).

2. I borrow the sense of "rhetoric" from Paul de Man, "Semiology and Rhetoric," *Diacritics* 3, no. 3 (1973): 27–33, especially the distinction between "semiology" and "rhetoric."

3. What do "citation" and "referencing" in an academic book mean, in a world in which there are various gadgets that can search texts within a few seconds or minutes, depending on techno-industrial connectivity? The very practice of verifiability

of sources is changing, and that needs to be acknowledged. These established practices of "authorization" (in print publication of academic work) need to be understood for what they have been, for we should not forget that beyond procedures of citation, there are court cases and judicial and economic processes of copyright and publication.

4. Gopal Guru and Sundar Sarukkai, *The Cracked Mirror: An Indian Debate on Experience and Theory* (New Delhi: Oxford University Press, 2012).

5. Ramnarayan Rawat and K. Satyanarayana, eds., *Dalit Studies* (Durham, N.C.: Duke University Press, 2016).

6. Those who are interested might want to read Aniket Jaaware, "Eating, and Eating with, the Dalit: A Reconsideration Touching upon Marathi Poetry," in *Modernism and After: Indian Poetry after 1947*, ed. K. Satchidanandan (New Delhi: Sahitya Akademi, 2001), 262–93.

7. Aniket Jaaware, "Destitute Literature," the First Phule Memorial Oration, University of Mumbai, December 2011, published as a pamphlet by the Phule-Ambedkar Chair, University of Mumbai.

1. TOUCH AND ITS ELEMENTS AND KINDS

1. See Aristotle, *The Poetics*, first line, almost any edition.

2. The most basic position in Immanuel Kant, *The Critique of Pure Reason*.

3. Martin Heidegger, *Being and Time*, trans. Joan Stambaugh (Albany: State University of New York Press, 1996), for the elaboration of "the ontical and the ontological priority of the question of being."

4. This is not to suggest that "scientific" research cannot tell us anything about touch. See the "Touch" section of the Scholarpedia: http://www.scholarpedia.org/article/Encyclopedia:Touch.

5. Maurice Merleau-Ponty, *The Phenomenology of Perception*, trans. Colin Smith (London: Routledge, 2002).

6. What would be the smallest measurable unit of touch or tactility? If there is to be a measure, it will have to combine rate and density.

7. This is a generalization; there possibly are art objects that only are meant to be experienced through touch.

8. For a variation on "shutting off" the sense of touch see chapter 2, "The Second Opposition: Literal and Figural Touch."

9. For example, the metaphor of the "body politic" in the European tradition or the notion of the *purusha* in the Hindu tradition.

10. It is customary to locate "caste" in endogamy and exogamy, in regulations of commensality and connubiality. Although these have been useful in sociology and anthropology, the argument here is that touchability/untouchability are fundamental to any understanding of caste, and the various anti-caste struggles that have

taken place in Maharashtra from Jotirao Phule onward, have had something to say about touchability/untouchability. The anti-caste struggles in Maharashtra could be said to have three basic phases: that of moral/ rational appeal (Phule, late nineteenth century), that of political and legal struggle (Ambedkar, early to mid-twentieth century), and that of "revolution" (Dalit Panther, 1960–90s).

11. See Thorstein Veblen, *The Theory of the Leisure Class*, http://www.gutenberg.org/files/833/833-h/833-h.htm, visited 06.01.2015.

12. Pierre Bourdieu, *Distinction: A Social Critique of the Judgement of Taste* (Cambridge, Mass.: Harvard University Press, 1984).

13. Walter Benjamin, "Capitalism as Religion," in *Selected Writings*, vol. 1: *1913–1926*, ed. Marcus Bullock and Michael W. Jennings (Cambridge, Mass.: Belknap Press of Harvard University Press, 1996), 288–91.

14. Louis Dumont, *Homo Hierarchicus: The Caste System and Its Implications*, trans. Mark Sainsbury, Louis Dumont, and Basia Gulati, rev. ed. (Chicago: University of Chicago Press, 1980).

15. Mary Douglas, *Purity and Danger: An Analysis of the Concepts of Pollution and Taboo* (London: Routledge, 2001). These and similar writings seem to assume that "purity" and "pollution" exist already and tend to describe the effects of these in society rather than attempt to understand *how* these come into being (their constitution) in society.

16. For a detailed discussion of animality and Heidegger, see D. F. Krell, *Daimon Life: Heidegger and Life-Philosophy* (Bloomington: Indiana University Press, 1992).

17. These are functions: mother, father, and so on. We are aware that there are many neonates who are not brought up by their biological parents.

18. See Trần Đức Thảo, *Investigation into the Origin of Language and Consciousness*, trans. Daniel J. Herman and Donald V. Morano (Boston: D. Reidel, 1986).

19. Jacques Lacan, "The Mirror Stage as Formative of the Function of the I as Revealed in Psychoanalytic Experience," in *Ecrits: A Selection*, trans. Alan Sheridan (London: Tavistock, 1977; reprinted Routledge, 1992).

20. Preferring the simpler and equally comprehensive explanation over the more complex ones. It could of course be argued that this is precisely what Freud did, locating the id in the inherited *soma* itself.

21. Michel Foucault, *Discipline and Punish: The Birth of the Prison*, trans. Alan Sheridan (Harmondsworth: Penguin Books, 1979).

22. M. M. Bakhtin, *The Dialogic Imagination: Four Essays*, ed. Michael Holquist, trans. Caryl Emerson and Michael Holquist (Austin: University of Texas Press, 1981).

23. Slavoj Žižek, *The Sublime Object of Ideology* (London: Verso, 1989).

24. Jacques Derrida, "Structure, Sign, and Play in the Discourse of the Human Sciences," in *The Structuralist Controversy: The Languages of Criticism and the Sciences of Man*, ed. R. Macksey and E. Donato (Baltimore: Johns Hopkins University Press, 1970).

25. Marcel Mauss, *The Gift: Forms and Functions of Exchange in Archaic Societies*, trans. Ian Cunnison (London: Cohen and West, 1954). This is a well-known and influential text, and major thinkers such as Lévi-Strauss and Jacques Derrida have responded, in their own ways. The influence of this text is too widespread to be listed here. My attempt here, however, is to focus on the "instantaneous" nature of the concept and experience of "exchange-value."

26. We are trying to see if there could be parallels between our understanding of allegory and touch, and allegory and reading as articulated by Paul de Man in his *Allegories of Reading: Figural Language in Rousseau, Nietzsche, Rilke, and Proust* (New Haven: Yale University Press, 1979).

2. TOUCH—AN A PRIORI APPROACH

1. Franz Stanzel, *A Theory of Narrative* (Cambridge: Cambridge Paperback Library, 1986), xvi.

2. Edith Stein, *On the Problem of Empathy*, trans. Waltraut Stein (The Hague: Martinus Nijhoff, 1964).

3. What we are trying to get at is the difference between familiarity and surprise, though contrary to what I have stated above, a willful touching can cause the experience of surprise. Thus, we are looking at degrees of surprise, and I suggest that when others touch us unexpectedly, that is the experience of being touched.

4. By "Romantic" I mean the mode of thinking which while privileging origins also renders them necessarily mysterious (e.g., various theories of poetic production, of which Wordsworth's "Preface to Lyrical Ballads" can be taken as a simple example).

5. Jacques Lacan, *Ecrits: A Selection*, trans. Alan Sheridan (London: Tavistock, 1977), translator's note, ix–x.

6. Hans-Georg Gadamer develops further Heidegger's notion of "fore-having" in his *Truth and Method*.

7. Viktor Shklovsky, "Art as Technique," in *Russian Formalist Criticism: Four Essays*, ed. Lee T. Lemon and Marion J. Reis (Lincoln: University of Nebraska Press, 1965).

8. I must note that in some variations of the ritual, water is poured over you by someone else. This could be seen as a substitution perhaps.

9. Richard Dawkins, *The Selfish Gene* (Oxford: Oxford University Press, 1976).

10. John Boswell, *The Kindness of Strangers: The Abandonment of Children in Western Europe from Late Antiquity to the Renaissance* (Chicago: University of Chicago Press, 1988).

11. Vishnubhat Godse, *Maza Pravas: 1857 cya Bandaci Hakikat* [My travels: The story of the 1857 mutiny], ed. Datto Vaman Potdar (Pune: Venus Prakashan, 1974). Translations mine.

12. The Carvaka or the Lokayata tradition is known to be radically empiricist, accepting only sensory experience as a valid means of knowledge, whereas the Vedic tradition is a ritualistic tradition that is "idealistic." Some other schools of thought accept as many as eight valid means of knowledge. For a translation see Madhavacharya, *Sarvadarsanhansamgraha; or, Review of the Different Systems of Indian Philosophy*, trans. E. B. Cowell and A. E. Gough (London: Truebner, 1882).

13. The tape does not have a production date.

14. A tool or a concept can be used for purposes it was not designed for (bricolage) only because that other undesigned use was potentially present in it (it is the work of history to discover those uses).

15. This has been the position of the Indian tradition of Leftism, including Indian Marxism, that caste is a matter of division of labor.

16. For a discussion of such and similar notions, see Edward Said, *Orientalism* (New York: Pantheon Books, 1978).

17. Kancha Ilaiah, *Why I Am Not a Hindu: A Sudra Critique of Hindutva Philosophy, Culture, and Political Economy* (1996; Calcutta: Samya, 2002) attempts what the title suggests but also tends to celebrate "shudra" culture without submitting it to a "critique," thereby weakening the philosophical force of the word *critique*.

18. See John McManners, ed., *The Oxford History of Christianity* (Oxford: Oxford University Press, 1990).

19. The second kind of metonymy is where the two realms of tropology and syllogistic logic actually converge: The cause for effect and the vice versa metonymies also often work as *inference*.

20. Things are not so simple though: From the shadow itself one can only infer that it is the shadow of a humanoid shape. We cannot *conclude* that it is a dalit's shadow. So metonymy too might need, sometimes, this moment of retrospection. Shadow, followed by a dalit, *therefore* a dalit's shadow, *therefore* contamination.

21. Walter Benjamin, "Analogy and Relationship," in *Selected Writings*, vol. 1: *1913–1926*, ed. Marcus Bullock and Michael W Jennings (Cambridge, Mass.: Belknap Press of Harvard University Press, 1996), 207–9.

22. We are acutely aware that the adjective "self-evident" is seriously questionable; what we mean by it is the fact that we do make the distinction between "good" and "bad," whatever the phenomenal content of that distinction. What is important to our argument is that the distinction is made, and that the content is determined by the entity that makes it. We do not mean anything empiricist by that adjective.

3. TOUCH IN ITS SOCIAL AND HISTORICAL ASPECTS I

1. In a certain sense, that distinction is empty of content, since we cannot find a nonmetaphysical society. But it is a necessary one, since it allows us to see ourselves as always already metaphysical in the literal and the metaphorical senses.

2. We are thinking of the various attempts in SF and F, at imagining "organic, harmonious" societies/communes, and so on. We could begin with Isaac Asimov's attempt at writing "I/we" or versions of these words, as well as the "Gaia" hypothesis, which attempts to think of the earth itself (and here, dramatically, it does not seem to be a Globe) as a spaceship, and so on. There have been attempts to imagine a harmonious and happy community, too numerous to be cited or even listed.

3. We use this word *samaaj* and this argument later as well. It might seem as if we were levering a lot of weight on a very weak fulcrum. However, is it not possible, in somebody else's language (in this case, *us*), that "society" only means divisions and distinctions and inequality?

4. We could have used the compound "nonhuman." We choose to use "inhuman" instead, since there has been and will be so much killing in the name of the human as well as the "divine," in various "societies."

5. It could be argued that one inherits "class" as well. But the stability (continuance) of that inheritance is much less than that of caste and/or gender, therefore we are treating it as a somewhat delayed inheritance in terms of chronology. One could get at this point by asking, "Does the neonate *have* class? If so, does it have class in the same way as it has caste and gender?"

6. *Shrimajjaiminipranite mimasadarshane [chaturthaadhyayamarabhya saptamaadhyaanto vibhagan] (bhattakumarilpranita tuptikayavyakhyasahita shabarabhashyopetah)* (Pune: Anandashram, 1984). There is a very long tradition of discussion of caste in ancient Sanskrit texts; we are treating this merely as a corroboration of our own thinking. For an argument about ancient Sanskrit texts, see chapter 5.

7. Mikhail Bakhtin, *The Dialogic Imagination: Four Essays*, ed. Michael Holquist, trans. Caryl Emerson and Michael Holquist (Austin: University of Texas Press, 1981).

8. By no means do we mean to suggest that there is no protest or dissent or sectarianism or resistance to power in societies of inheritance. The point is that these do not really lead to a fundamental transformation of society or religion or religiosity.

9. Northrop Frye, *The Anatomy of Criticism* (Princeton, N.J.: Princeton University Press, 1957).

10. The importance of the distinction between analytical and synthetic judgements, especially after Kant's elaboration of it in *The Critique of Pure Reason*, cannot be overemphasized.

11. *Bhakti* poetry was a movement in religion and poetry that was very popular from about the twelfth century onward in various parts of India. Many of the poets were also called saints. *Bhakti* translates, very roughly, as devotion. Although it is true that many *bhakti* authors criticized caste, refused to practice or validate it, historically this criticism had its own limitations and did not become common to

all people. We suggest here that it is because of their religiosity/spirituality (though different from institutional religion) that these movements failed to make much practical impact. A lot has been written about this, which we cannot go into here.

12. See Diagram 1 in chapter 2.

13. A clear formulation of this notion is found in Martin Heidegger, *Being and Time*.

14. What we mean is that the statement assumes a didactic and critical metaso-cial position, unavailable to the addressee, who is supposed to receive this enlight-ening sentence as an incisive comment on his or her social position, and understand it in all humility.

15. The *Kaamshastras* (science/systematic study/manuals of erotics) categorize women into "types" based on some physical attributes. A *shankhini*, for example, is woman who has a narrow neck like a conch. *Padmini* would mean "lotus type."

16. Upper-caste men typically wear a thickly plaited thread from the left shoul-der that turns around the right of the abdomen. In principle this thread is never taken off. For details in the tradition, see http://www.vedpradip.com/articlecontent.php?aid=184&linkid=1&catid=0&subcatid=0, visited June 12, 2016

17. This changed for the worse in the 2002 riots in Gujarat.

18. A *yajna* is usually a sacrificial ritual aimed at achieving a certain "fruit," or "result." This particular ritual is meant to get the performer a son.

19. Charles Malamoud, *Cooking the World: Ritual and Thought in Ancient India*, trans. David White (Delhi: Oxford University Press, 1996).

20. Xinru Liu, *Silk and Religion: An Exploration of Material Life and the Thought of the People, AD 600–1200* (Delhi: Oxford University Press, 1998).

21. Jotirao Phule, *Shetkaryacha Asud*, in *Mahatma Phule Samagra Vangmay*, ed. Y. D. Phadke (Mumbai: Maharashtra Rajya Sahitya Ani Sanskruti Mandal, 1991). See also "The Cultivator's Whip-Cord," trans. Aniket Jaaware, in *Selected Writings of Jotirao Phule*, ed. G. P. Deshande (Delhi: Leftword, 2002).

22. These texts were the main sources for a legal settlement and legal ap-peals and suits in the Hindu tradition in Maharashtra until the late nineteenth century.

23. Seema Alavi, *The Sepoys and the Company: Tradition and Transition in Northern India, 1770–1830* (Delhi: Oxford University Press, 1995).

24. The Mahar Regiment would have had people of the Mahar caste.

25. *Reservation* indicates something like affirmative action in the United States. The central and state governments ensure that a certain number of jobs (graded according to caste) are "reserved" for "lower-caste" aspirants. This is a matter of bitter debate in India.

26. Walter Benjamin, "Capitalism as Religion," in *Selected Writings*, vol. 1: *1913-1926*, ed. Marcus Bullock and Michael W. Jennings (Cambridge, Mass.: Belknap Press of Harvard University Press, 1996), 288–91.

4. TOUCH IN ITS SOCIAL AND HISTORICAL ASPECTS II

1. The methods and ways of reading official hegemonic documents to "reconstruct" subaltern histories as developed mainly by the Subaltern Studies Collective could possibly be replicated, but that line of thought would be different from the one we are taking.

2. The point is that cultural rememoration, personal affect, and similar things interfere more in the facts of the past than of the present—it is *possible* to *pretend* that the present is "immediately empirical," even if further investigation reveals that it is not.

3. For example, Le Roy Ladurie, *Montaillou: Cathars and Catholics in a French Village, 1294–1324* (French ed., 1978), trans. Barbara Bray (London: Penguin, 1990).

4. This is the contradiction that we have to struggle with: If something or other is a "whole," it has to be complete, whereas "society" is never ever really complete, it has an "open-ended" temporality (Bakhtin), one in which anything can happen at the next moment, completely unexpectedly.

5. Privilege and unprivilege are inherited, and social transformation therefore might mean giving up *both*.

6. The metaphorical/allegorical serpent in the body.

7. B. R. Ambedkar, *Who Were the Shudras?* in *Dr Babasaheb Ambedkar Writings and Speeches*, vol. 7 (Bombay: Education Department Government of Maharashtra, 1990).

8. Trần Đức Thảo, *Phenomenology and Dialectical Materialism*, ed. Robert S. Cohen, trans. Daniel J. Herman and Donald V. Morano (Boston: D. Reidel, 1986).

9. It seems to us that this is where the resolution of the contradiction is not accepted within the tradition: something different from the Hegelian resolution of it, by having an always already different—and higher level—contradiction to resolve. But these are difficult ideas, and we are not sure we have it right.

10. This does *not* mean that dalits do not practice touchability/untouchability within the caste subdivisions.

11. For a discussion of Greimas's concepts, see Ronald Schleifer, *A. J. Greimas and the Nature of Meaning: Linguistics, Semiotics, and Discourse Theory* (London: Croom Helm, 1987).

12. See, inter alia, Marcel Mauss, *The Gift: The Forms and Functions of Exchange in Archaic Societies*, trans. Ian Cunnison (London: Cohen and West, 1954), and Jacques Derrida, *Given Time I: Counterfeit Money*, trans. Peggy Kamuf (Chicago: University of Chicago Press, 2017).

13. We have already pointed out this feature with reference to metaphor; now it turns to be true for metonymy as well.

14. Ram Bapat (1931–2012) retired as professor from the Department of Politics and Public Administration, University of Pune. He was a very important intellectual figure in Maharashtra and elsewhere.

15. Unfortunately, Professor Bapat passed away before we could discuss this further and get to the resources on which he based his observation.

16. B. B. Acharya and M. V. Shingane, *Mumbaicha Vruttant* [An account of/report on Bombay] (Mumbai: Nirnaysagar Press, 1889).

17. A mark applied on the forehead.

18. We are partly drawing on Gilbert Simondon, "Technical Mentality," trans. Arne De Boever, *Parrhesia*, no. 7 (2009): 17–27.

19. By "special cultural encoding" we mean the emphasis on emancipation as necessarily modern.

20. We have already mentioned cultural re-memoration earlier.

21. Pierre Bourdieu, *Distinction: A Social Critique of the Judgement of Taste*, trans. Richard Nice (Cambridge, Mass.: Harvard University Press, 1984).

5. TOUCH AND TEXTS: ANCIENT AND MODERN

1. Jacques Derrida, "Biodegradables: Seven Diary Fragments," *Critical Inquiry* 15, no. 4 (1989): 812–73.

2. This is not the case with modern dalit literature as a narrative of suffering: Only those who survive the suffering *can* narrate/narrativize it.

3. Brian Stock, *The Implications of Literacy: Written Language and Models of Interpretation in the Eleventh and Twelfth Centuries* (Princeton, N.J.: Princeton University Press, 1983).

4. "The ideas of the ruling class are in every epoch the ruling ideas, i.e. the class which is the ruling material force of society, is at the same time its ruling intellectual force." Karl Marx, *The German Ideology*, https://www.marxists.org/archive/marx/works/1845/german-ideology/cho1b.htm, visited August 14, 2016.

5. Hans-Georg Gadamer, *Reason in the Age of Science*, trans. Frederick G. Laurence (Cambridge, Mass.: MIT Press, 1982).

6. Those interested in the detail on this should see the text known in the tradition as Parthasarathy Mishra's *Shastradeepika*, ed. Lakshman Shastri Dravid (Benares: Choukhamba Oriental Series, 1916).

7. http://www.merriam-webster.com/dictionary/expressio%20unius%20est%20exclusio%20alterius, visited August 14, 2016. Roughly, "the expression of one is the exclusion of the other."

8. Aniket Jaaware, "Eating, and Eating with, the Dalit: A Reconsideration Touching upon Marathi Poetry," in *Indian Poetry after 1947*, ed. K. Satchidanandan (New Delhi: Sahitya Akademi, 2001), 262–93.

9. Walter Ong, *Orality and Literary: The Technologizing of the Word* (London: Methuen, 1982).

10. We admit that it is somewhat anachronistic to call Phule a "dalit"; however, he has been so deeply influential and popular in the dalit tradition that it seems appropriate to call him an early dalit thinker.

11. "I salute you / I bow to you." Unlike "zohaar," this phatic expression is not marked by caste explicitly.

12. For a relatively recent treatment of related issues and the saint-poet Kabir, see Milind Wakankar, *Subalternity and Religion: The Prehistory of Dalit Empowerment in South Asia* (London: Routledge, 2010). Our treatment is less respectful of saint-poets and the ethical efficacy of spirituality.

13. Jotirao Phule, *Selected Writings of Jotirao Phule*, ed. G. P. Deshpande (New Delhi: LeftWord Books 2002), 141–44.

14. With hindsight, we are now in a position to suggest that dalit politics and life would have been different had atheism been chosen at that point in time, instead of Buddhism. The issue has many facets and explanations. In many personal discussions a point of view has been offered, which mostly amounts to the cynical explanation that Ambedkar converted to another religion because "the people" would not become atheists *en masse*. This not acceptable to me, since it relegates "the people" to a not very intelligent mass. For an argument that Ambedkar's interpretation of Buddha's *dhamma* was not a religious one, see Pradeep Gokhale, "Ambedkar and Modern Buddhism: Continuity and Discontinuity," Buddha Jayanti Lecture on 28th December 2013 in the 88th session of Indian Philosophical Congress held at Madurai.

15. B. R. Ambedkar, *Who Were the Shudras?* in *Writings and Speeches* (Mumbai: Government of India), 7:32.

16. Ibid., 187–88.

17. Elizabeth Eisenstein, *The Printing Revolution in Early Modern Europe* (Cambridge: Cambridge University Press, 1983).

18. Ilya Prigogine and Isabelle Stengers, *Order out of Chaos: Man's New Dialogue with Nature* (Boulder, Colo.: Shambhala, 1984).

19. Gabriel Tarde, *Laws of Imitation*, https://archive.org/details/lawsimitation01 tardgoog.

20. Aniket Jaaware, "Destitute Literature," First Mahatma Jyotirao Phule Oration (Mumbai: Mahatma Jyotirao Phule and Dr Babasaheb Ambedkar Chair, 2012).

21. Jindřich Toman, *The Magic of a Common Language: Jakobson, Mathesius, Trubetzkoy, and the Prague Linguistic Circle* (Cambridge, Mass.: MIT Press, 1995). For Kartsevsky, see. 120, 136; for Trubetzkoy, 146–47.

22. For Al-Biruni see http://www.columbia.edu/cu/lweb/digital/collections /cul/texts/ldpd_5949073_001/index.html.

23. For Abbe Dubois see https://archive.org/stream/hindumannerscust1906dubo/hindumannerscust1906dubo_djvu.txt.

24. For a contextualized description of Ziegenbalg, see Will Sweetman, "The Prehistory of Orientalism: Colonialism and the Textual Basis for Bartholomaus Ziegenbalg's Account of Hinduism," *New Zealand Journal of Asian Studies* 6, no. 2 (2004): 12–38, http://www.otago.ac.nz/religiousstudies/staff/articles/prehistory.pdf.

25. Sitaram Pandey, *From Sepoy to Subedar: Being the Life and Adventures of Subedar Sitaram, a Native Officer of the Bengal Army, Written and Related by Himself*, trans. Lieutenant-Colonel Norgate, ed. James Lunt (Delhi: Vikas Publications, 1970; first published Bengal Staff Corps, Lahore, 1873), 167–69.

26. We are repeating here the argument made in "Destitute Literature."

6. (UN)TOUCHABILITY OF THINGS AND PEOPLE

1. Large parts of the argument that follows draw on Aniket Jaaware, "Destitute Literature," First Mahatma Jyotirao Phule Oration (Mumbai: Mahatma Jyotirao Phule and Dr Babasaheb Ambedkar Chair, 2012).

2. This is again a summary of our argument in "Destitute Literature."

3. It is possible to argue that Heidegger's discussion of "being-with" and "being-in-the-world" is his way of taking into account the sociality of existence. But this is not really made explicit. Therefore, this attempt.

4. Among others, Michel Foucault, *Power/ Knowledge: Selected Interviews and Other Writings, 1972–1977*, ed. Colin Gordon (New York: Pantheon Books, 1980).

5. We believe that *conspiracy* is the collective noun for bureaucrats; another one is *shuffle*.

6. We have earlier said that Phule has a lot to say about religion. What he did was to construct another, with a creator God, and with completely different rituals and prayers, a religion called "truth-seeker" (*satya-shodhak*). For example, there is a prayer, in Sanskrit, which worships and pays obeisance to a long list of various creatures, beginning with insects.

7. See, inter alia, Parthasarathy Mishra, *Shastradeepika*, ed. Lakshman Shastri Dravid (Benares: Choukhamba Oriental Series, 1916).

8. J. L. Austin, *How to Do Things with Words* (Oxford: Clarendon Press, 1962).

9. William Herschel, *The Origin of Finger-Printing*, http://www.gutenberg.org/files/34859/34859-h/34859-h.htm.

10. Francis Galton, *Finger Prints*, http://galton.org/books/finger-prints/galton-1892-fingerprints-1up.pdf.

11. Henry Faulds, *Guide to Finger-print Identification,* http://galton.org/fingerprints/books/faulds/faulds-1905-guide-1up.pdf.

12. Gottlob Frege, "On Sense and Reference," trans. Max Black, *Philosophical Review* 57, no. 3 (1948): 209–30.

13. These developments are documented in Richard Rorty, *The Linguistic Turn: Essays in Philosophical Method* (Chicago: University of Chicago Press, 1967, with additional material, 1992). We give them here in miniature.

14. For a recent introduction to and discussion of the Charvaka school of thought, see Pradeep P. Gokhale, *Lokayata/ Carvaka: A Philosophical Inquiry* (Delhi: Oxford University Press, 2015).

15. See, for example, Jacques Derrida, "The Ends of Man," in *Margins of Philosophy*, trans. Alan Bass (Chicago: University of Chicago Press, 1986), 109–36.

16. Claude Lévi-Strauss, "The Structural Study of Myth," *Journal of American Folklore* 68, no. 270 (1955): 428–44. Web.

17. This might be yet another way to understand David Hume's well-known skepticism about "causality": He seems to suggest that merely because two things always go together (spatial and temporal concomitance), it does not follow that one thing is *caused* by the other (inference). See William Edward Morris and Charlotte R. Brown, "David Hume," in *The Stanford Encyclopedia of Philosophy* (Spring 2016 edition), ed. Edward N. Zalta, http://plato.stanford.edu/archives/spr2016/entries/hume, visited June 18, 2016.

18. Marshall Sahlins, *Stone Age Economics* (New York: de Gruyter, 1972).

19. Perhaps she uses a Dvorak, or some other keyboard, that does not make a difference to the description.

7. SOCIETY, SOCIALITY, SOCIABILITY

1. What we have in mind is the ability-as-possibility, which could even be treated as a transcendental condition of possibility for something like social behavior. This has nothing to do really with being nice to other people, or, for that matter, with altruism.

2. Almost uncannily, the metaphor of shutting doors occurs in B. R. Ambedkar's discussion of the origin of caste, in his 1916 essay, "Castes in India: Their Mechanism, Genesis, and Development."

3. Perhaps it is possible to understand Heidegger's notion of "being-with" not merely as being a structural part of an "Analytic of Dasein" but also as an everyday task of inventive sociability in its everydayness.

4. Stefanie Trojan, http://www.stefanietrojan.de.

5. It is possible to understand this *process* in a Hegelian manner: The future state will give up and yet incorporate the past state. See G. W. F. Hegel, *Phenomenology of Spirit*, trans. A. V. Miller, (Oxford: Oxford University Press, 1977), 2, 68.

6. Michel Foucault, *Society Must Be Defended: Lectures at the Collège de France, 1975–76*, trans. David Macey, ed. Mauro Bertani and Alessandro Fontana (New York: Picador, 2003), 60ff.; see also Achille Mbembe's treatment of Foucault's notion of race wars, in "Necropolitics," trans. Libby Meientjes, *Public Culture* 15, no. 1 (2003): 11–40.

7. Traverse: early 14 c., "pass across, over, or through," from Old French traverser "to cross, place across" (11 c.), from Vulgar Latin *traversare, from Latin transversare "to cross, throw across," from Latin transversus "turn across" (see transverse). As an adjective from early 15 c. Related: Traversed; traversing. http://etymonline.com/index.php?term= traverse&allowed_in_frame=0.

8. Inasmuch as "re-sistance" means to stand firm again and again, the premature moral satisfaction that those who claim to be resistant should caution us. To stand firm again would mean that the earlier standing firm (resistance) was not effective. If we are having to resist today, then earlier resistance must have failed. If we have to resist tomorrow, by that time, today's resistance must inevitably fail. It seems to me that resistance, although morally a very comforting idea, might not be as useful for a practical politics aimed at social change.

9. Kierkegaard's aphorism: "Alas that the doors of Destiny open outwards; had they opened inwards, we could push them open." Another version: "The door to happiness opens outwards. Anyone who tries to push this door open thereby causes it to close still more."

10. This power-to-kill need not be contained "inside" the notion of sovereign power as its subset, again, see Mbembe, "Necropolitics."

11. William Wordsworth, "No motion has she now, no force / She neither hears not sees / Rolled around in earth's diurnal course."

12. John Donne, "At the round earth's imagin'd corners, blow / Your trumpets, angels, and arise, arise / From death, you numberless infinities / Of souls, and to your scatter'd bodies go."

13. We should remember that Ambedkar himself had offered to burn the constitution in 1953.

14. https://en.wikipedia.org/wiki/Burakumin, visited June 18, 2016. See also https://socialistworker.org/2012/04/02/japans-untouchable-workers.

8. RECAPITULATION WITH VARIATIONS

1. We have not until now discussed the word *caste* itself. There have been arguments that the word is of foreign origin and in any case does not "capture" the Indian reality of social relations of inequality. We are of the opinion that the word has done a lot of work in our intellectual history, and we should let it do the work it is doing. Sanskrit or Sanskritized words such as *varna* or *jaati* do not do the same work. It is not logical to assume that "Indian" words will necessarily "capture" Indian reality better.

2. *Alamkara* is the Sanskrit word for "ornament"; a study of ornamentation is called *alamkara-shastra*.

3. By apophantic we mean self-evident "analytic judgments," where the predicate is logically or empirically a part of the subject.

4. This is the Sanskrit/ Marathi word for "device."

5. Gregory Bateson, *Naven: A Survey of the Problems Suggested by a Composite Picture of the Culture of a New Guinea Tribe Drawn from Three Points of View* (Stanford: Stanford University Press, 1936).

6. http://www.eoht.info/page/Bifurcation, visited March 22, 2016, and http://firstmonday.org/ojs/index.php/fm/article/view/687/597, visited March 22, 2016.

7. These are the electoral symbols of various political parties in Maharashtra and India.

8. This poem by Emily Dickinson is too well known to give a reference.

9. These are the first three words of the Indian Constitution, without the question mark.

BIBLIOGRAPHY

Aarsleff, Hans. *The Study of Language in England, 1780–1860*. Princeton, N.J.:
Princeton University Press, 1967.

Acharya B. B., and M. V. Shingane. *Mumbaicha Vruttant*. Mumbai: Nirnaysagar
Press, 1889. [An account of Mumbai.]

Adorno, Theodor. *Negative Dialectics*. Translated by E. B. Ashton. London:
Continuum, 2007.

Alavi, Seema. *The Sepoys and the Company: Tradition and Transition in Northern
India, 1770–1830*. Delhi: Oxford University Press, 1995.

Althusser, Louis. "Ideology and Ideological State Apparatuses (Notes Towards an
Investigation). In *Essays on Ideology*. London: Verso, 1984.

Althusser, Louis, and Etienne Balibar. *Reading Capital*. Translated by Ben
Brewster. London: Verso, 1979.

Ambedkar, B. R. *Annihilation of Caste*. In *Dr Babasaheb Ambedkar Writings and
Speeches*, 1:25–96. Bombay: Government of Maharashtra, 1979.

———. "Castes in India: Their Mechanism, Genesis, and Development." In *Dr
Babasaheb Ambedkar Writings and Speeches*, vol. 1. Mumbai: Government of
Maharashtra, 1979.

———. *Who Were the Shudras? How They Came to Be the Fourth Varna in the
Indo-Aryan Society*. In *Dr Babasaheb Ambedkar Writings and Speeches*, vol. 7.
Mumbai: Government of Maharashtra, 1990.

Aristotle. *Metaphysics*. Translated by W. D. Ross. Adelaide: University of Adelaide
Library, 2000.

———. *Nicomachean Ethics*. Translated by W. D. Ross and Lesley Brown. Oxford:
Oxford University Press, 2009.

———. *Physics*. Translated by W. D. Ross. Oxford: Clarendon Press, 1960.

———. *Poetics*. Translated by M. Heath; edited with introduction and annotations
by A. Das Gupta. Contains supplementary essays, Penguin Study Editions.
Delhi: Pearson Books, 2007.

Austin, J. L. *How to Do Things with Words: The William James Lectures delivered at Harvard University in 1955.* Edited by J. O. Urmson. Oxford: Clarendon Press, 1962.

Bagul, Baburao. *Maran Swasta Hot Ahe* [Death is becoming cheaper, short stories]. Mumbai: Lokvangmaya Griha, 1969.

———. *Sood* [Revenge, a novella]. Mumbai: Lokvangmaya Griha, 1970.

Bakhtin, Mikhail M. *The Dialogic Imagination: Four Essays.* Edited by Michael Holquist; translated by Caryl Emerson and Michael Holquist. Austin: University of Texas Press, 1981.

———. *Speech Genres and Other Late Essays.* Translated by Caryl Emerson and Michael Holquist. Austin: University of Texas Press, 1986.

Balgangadhara, S. N. *"The Heathen in His Blindness": Asia, the West and the Dynamic of Religion.* Leiden: E. J. Brill, 1994.

Barthes, Roland. "An Introduction to the Structural Analysis of Narrative." In *Image-Music-Text.* Translated by Stephen Heath. London: Fontana, 1977.

Bateson, Gregory. *Naven: A Survey of the Problems Suggested by a Composite Picture of the Culture of a New Guinea Tribe Drawn from Three Points of View.* Stanford: Stanford University Press, 1936.

———. *Steps to an Ecology of Mind: Collected Essays in Anthropology, Psychiatry, Evolution, and Epistemology.* Chicago: University of Chicago Press, 1972.

Benjamin, Walter. *Selected Writings*, vols. 1–4. Edited by Marcus Paul Bullock, Michael William Jennings, Howard Eiland, and Gary Smith; translated by Rodney Livingstone. Cambridge, Mass.: Belknap Press of Harvard University Press, 1996–2003.

Bhatta, Kumaril. *Tupteeka.* (*Shrimajjaiminipranite mimasadarshane [chaturthaadhyayamarabhya saptamaadhyaanto vibhagan] (bhattakumarilpranita tuptikayavyakhyasahita shabarabhashyopetah).* Pune: Anandashram, 1984. [Jaimini's *mimamsa-darshana* from the fourth to the seventh chapters, with a commentary by Kumaril Bhatta called *tuptika* and Shabara's commentary.]

Boswell, John. *The Kindness of Strangers: The Abandonment of Children in Western Europe from Late Antiquity to the Renaissance.* Chicago: University of Chicago Press, 1988.

Bourdieu, Pierre. *Distinction: A Social Critique of the Judgement of Taste.* Translated by Richard Nice. Cambridge, Mass.: Harvard University Press, 1984.

———. *The Logic of Practice.* Translated by Richard Nice. Stanford: Stanford University Press, 1990.

Burton, Robert. *Anatomy of Melancholy.* Edited by Thomas C. Faulkner, Nicolas K. Kiessling, and Rhonda L. Blair. 3 vols. Oxford: Clarendon Press, 1989.

Dawkins, Richard. *The Blind Watchmaker.* New York: W. W. Norton, 1986.

——. *The Extended Phenotype.* Oxford: Oxford University Press, 1982.

——. *The Selfish Gene.* Oxford: Oxford University Press, 1976.

de Man, Paul. *Allegories of Reading: Figural Language in Rousseau, Nietzsche, Rilke, and Proust.* New: Yale University Press, 1979.

——. "Semiology and Rhetoric." *Diacritics* 3, no. 3 (1973): 27–33.

Derrida, Jacques. "Biodegradables: Seven Diary Fragments." *Critical Inquiry* 15, no. 4 (1989): 812–73.

——. "The Ends of Man." In *Margins of Philosophy*, translated by Alan Bass, 109–36. Chicago: University of Chicago Press, 1982.

——. *Margins of Philosophy.* Translated by Alan Bass. Chicago: University of Chicago Press, 1982.

——. *Of Grammatology.* Translated by Gayatri Chakravorty Spivak. Delhi: Motilal Banarasidas Publishers, 2002.

——. *On Touching: Jean-Luc Nancy.* Translated by Christine Irizarry. Stanford: Stanford University Press, 2005.

——. "Structure, Sign and Play in the Discourse of the Human Sciences." In *The Structuralist Controversy: The Languages of Criticism and the Sciences of Man*, edited by R. Macksey and E. Donato. Baltimore: Johns Hopkins University Press, 1970.

——. *Writing and Difference.* Translated by Alan Bass. London: Routledge & Kegan Paul, 1978.

Dickinson, Emily. "I'm Nobody!" In *Poems, Series 2*, edited by Mabel Loomis Todd and Thomas Wentworth Higginson. Boston: Robert Brothers, 1891.

Dirks, Nicholas B. *Castes of Mind: Colonialism and the Making of Modern India.* Princeton, N.J.: Princeton University Press, 2001.

Douglas, Mary. *Purity and Danger: An Analysis of the Concepts of Pollution and Taboo.* London: Routledge, 2001.

Dumont, Louis. *Homo Hierarchicus: The Caste System and Its Implications.* Translated by Mark Sainsbury, Louis Dumont, and Basia Gulati. Chicago: University of Chicago Press, 1980.

Eisenstein, Elizabeth. *The Printing Revolution in Early Modern Europe.* Cambridge: Cambridge University Press, 1983.

Faulds, Henry. *Guide to Finger-print Identification.* http://galton.org/fingerprints/books/faulds/faulds-1905-guide-1up.pdf.

——. *A Manual of Practical Dactylography: A Work for the Use of Students of the Finger-print Method of Identification.* London: Police Review, 1923.

Foucault, Michel. *Discipline and Punish: The Birth of the Prison.* Translated by Alan Sheridan. Harmondsworth: Penguin Books, 1979.

——. *Power/ Knowledge: Selected Interviews and Other Writings, 1972–1977.* Edited and translated by Colin Gordon. New York: Pantheon Books, 1980.

————. *Society Must Be Defended: Lectures at the Collège de France, 1975–76.* Translated by David Macey, edited by Mauro Bertani and Alessandro Fontana. New York: Picador, 2003.

Frege, Gottlob. "On Sense and Reference." Translated by Max Black. *Philosophical Review* 57, no. 3 (1948): 209–30.

Freud, Sigmund. *The Interpretation of Dreams.* Translated by James Strachey. Pelican Freud Library, vol. 4. Harmondsworth: Penguin, 1949.

————. *An Outline of Psychoanalysis.* Translated by James Strachey. London: Hogarth Press and the Institute of Psycho-Analysis, 1949.

Frye, Northrop. *The Anatomy of Criticism.* Princeton, N.J.: Princeton University Press, 1957.

Gadamer, Hans-Georg. *Reason in the Age of Science.* Translated by Frederick G Laurence. Cambridge, Mass.: MIT Press, 1982.

————. *Truth and Method.* Translated by Joel Weinsheimer and Donald G. Marshall. New York: Continuum, 2004.

Galton, Francis. *Finger Prints,* http://galton.org/books/finger-prints/galton-1892 -fingerprints-1up.pdf.

Ghurye, Govind Sadashiv. *Caste and Race in India.* Bombay: Popular Prakashan, 2008.

Godse, Vishnubhat. *Maza Pravas: 1857 chya Bandaci Hakikat.* Edited by Datto Vaman Potdar. Pune: Venus Prakashan, 1974.

Gokhale, Pradeep P. *Lokayata/ Carvaka: A Philosophical Inquiry.* Delhi: Oxford University Press, 2015.

Gramsci, Antonio. *Prison Notebooks,* vols. 1–4. Edited and translated by Joseph A. Buttigieg and Antonio Callari. New York: Columbia University Press, 2011.

Guru, Gopal, and Sundar Sarukkai. *The Cracked Mirror: An Indian Debate on Experience and Theory.* New Delhi: Oxford University Press, 2012.

Hegel, G. W. F. *Phenomenology of Spirit.* Translated by A. V. Miller. Oxford: Oxford University Press, 1977.

Heidegger, Martin. *Being and Time: A Translation of Sein und Zeit.* Translated by Joan Stambaugh. Albany: State University of New York Press, 1996.

Henry, Edward Richard, and Sir Bart. *Classification and Uses of Finger Prints.* London: Home Office, 1905.

Herschel, William James. *The Origin of Finger-printing.* New York: H. Milford, 1916.

Husserl, Edmund. *Cartesian Meditations: An Introduction to Phenomenology.* Translated by Dorion Cairns. The Hague: M. Nijhoff, 1960.

————. *Logical Investigations,* vols. 1–2. Translated by J. N. Findlay. London: Routledge, 2001.

Ilaiah, Kancha. *Why I Am Not a Hindu: A Sudra Critique of Hindutva Philosophy, Culture, and Political Economy.* 1996. Calcutta: Samya, 2002

Ingarden, Roman. *The Cognition of the Literary Work of Art*. Translated by Ruth Ann Crowley and Kenneth R Olson. Evanston, Ill.: Northwestern University Press, 1973.

———. *The Literary Work of Art: An Investigation on the Borderlines of Ontology, Logic, and Theory of Literature*. Translated by George G. Grabowicz. Evanston, Ill.: Northwestern University Press, 1973.

Jaaware, Aniket. "Destitute Literature." First Mahatma Jyotirao Phule Oration. Mumbai: Mahatma Jyotirao Phule and Dr. Babasaheb Ambedkar Chair, 2012.

———. "Eating, and Eating with, the Dalit: A Reconsideration Touching upon Marathi Poetry." In *Indian Poetry after 1947*, ed. K. Satchidanandan, 262–93. New Delhi: Sahitya Akademi, 2001.

Jakobson, Roman. *Language in Literature*. Edited by Krystyna Pomorska and Stephen Rudy. Cambridge, Mass.: Belknap Press of Harvard University Press, 1987.

———. "Linguistics and Poetics." In *Style in Language,* edited by Thomas Sebeok, 350–77. Cambridge, Mass.: MIT Press, 1960.

Jennings, Humphrey. *Pandemonium: The Coming of the Machine as Seen by Contemporary Observers, 1660–1886*. New York: Free Press, 1985.

Kant, Immanuel. *The Critique of Judgement*. Translated with introduction and notes by J. H. Bernard. New York: Dover, 2005.

———. *The Critique of Pure Reason*. Translated by Norman Kemp-Smith. London: Macmillan, 1933.

Karve, Irawati. *Maharashtra, Land and Its People*. Vol. 60. Maharashtra: Directorate of Government Printing, Stationery, and Publications, 1968.

———. *Yuganta: The End of an Epoch*. New Delhi: Orient Longman, 2006.

Ketkar, S. V., and A. K. Datta. *Brahman-Kanya*. New Delhi: National Book Trust, 1945.

Kharat, Shankarrao. *Taral-Antaral*. Pune: Continental Prakashan, 1981.

Kierkegaard, Søren. *Either/Or: A Fragment of Life*. Translated by A. Hannay. London: Penguin, 2004.

———. *Kierkegaard's Writings, VI: Fear and Trembling/Repetition*. Edited by Edna H. Hong and Howard V. Hong. Princeton, N.J.: Princeton University Press, 2013.

Krell, David Farrell. *Daimon Life: Heidegger and Life-Philosophy*. Bloomington: Indiana University Press, 1992.

Lacan, Jacques. *Ecrits: A Selection*. Translated by Alan Sheridan. London: Tavistock, 1977.

Ladurie, Le Roy. *Montaillou: Cathars and Catholics in a French Village, 1294–1324*. Translated by Barbara Bray. London: Penguin, 1990.

Lakoff, George, and Mark Johnson. *Metaphors We Live By*. Chicago: University of Chicago Press, 1980.

Le Guin, Ursula. *The Dispossessed: An Ambiguous Utopia*. New York: Harper & Row, 1974.

———. *Gifts*. Orlando, Fla.: Harcourt, 2004.

———. *Powers*. Orlando, Fla.: Harcourt, 2007.

———. *Voices*. Orlando, Fla.: Harcourt, 2008.

Lévi-Strauss, Claude. *Race and History*. Paris: UNESCO, 1952.

———. *The Raw and the Cooked: Introduction to a Science of Mythology*. Translated by John Weightman and Doreen Weightman. New York: Octagon Books, 1970.

———. "The Structural Study of Myth." *Journal of American Folklore* 68, no. 270 (1955): 428–44.

Madhvacharya. *Sarvadarsanhansamgraha; or, Review of the Different Systems of Indian Philosophy*. Translated by E. B. Cowell and A. E. Gough. London: Truebner, 1882.

Malamoud, Charles. *Cooking the World: Ritual and Thought in Ancient India*. Translated by David White. Delhi: Oxford University Press, 1996.

Manu. *Manusmrti*. Edited by Ramchandra Varma Shastri. New Delhi: Vidya Vihar, 1997.

Marx, Karl, et al. *Capital*. Vol. 1: *A Critical Analysis of Capitalist Production*. Edited by Dona Torr. London: Allen & Unwin, 1938.

Marx, Karl, and Friedrich Engels. *The Economic and Philosophic Manuscripts of 1844 and the Communist Manifesto*. Translated by Martin Milligan. New York: Dover, 2007.

Mauss, Marcel. *The Gift: The Forms and Functions of Exchange in Archaic Societies*. Translated by Ian Cunnison. London: Cohen and West, 1954.

Mbembe, Achille. "Necropolitics." Translated by Libby Meientjes. *Public Culture* 15, no. 1 (2003): 11–40.

McManners, John, ed., *The Oxford History of Christianity*. Oxford: Oxford University Press, 1990.

Merleau-Ponty, Maurice. *The Phenomenology of Perception*. Translated by Colin Smith. London: Routledge, 2002.

Miller, Walter M., Jr. *A Canticle for Leibowitz*. New York: Bantam Books, 1997.

Mishra, Parthasarathy. *Shastradeepika*. Edited by Lakshman Shastri Dravid. Benares: Choukhamba Oriental Series, 1916.

Nietzsche, Friedrich. *Twilight of the Idols, and the Anti-Christ*. Translated and edited, with a commentary, by R. J. Hollingdale. Harmondsworth: Penguin Books, 1979.

Ong, Walter J. *Orality and Literacy: The Technologizing of the Word*. London: Methuen, 1982.

Oshima, Nagisa, dir. *The Ceremony*. Written by Nagisa Oshima, Mamoru Sasaki, and Tsutomu Tamura. Japan: Art Theatre Guild, 1971.

Padmanji, Baba. *Yamuna-Paryatan.* Bombay: T. Graham, 1857.

Pandey, Sitaram. *From Sepoy to Subedar: Being the Life and Adventures of Subedar Sitaram, a Native Officer of the Bengal Army, Written and Related by Himself.* Translated by Lieutenant-Colonel Norgate. Edited by James Lunt. Lahore: Bengal Staff Corps,1873. Reprint: Delhi: Vikas Publications, 1970.

Patil, Sharad. *Dasa-Sudra Slavery.* Bombay: Allied Publishers, 1978.

Phule, Jotirao. *Selected Writings of Jotirao Phule.* Edited, with annotations and introduction, by G. P. Deshpande. New Delhi: LeftWord Books, 2002.

Prigogine, Ilya, and Isabelle Stengers. *Order out of Chaos: Man's New Dialogue with Nature.* Boulder, Colo.: Shambhala, 1984.

Pullman, Philip. *His Dark Materials Trilogy.* London: Scholastic, 1995–2000.

Rao, Anupama. *The Caste Question: Dalits and the Politics of Modern India.* Berkeley: University of California Press, 2009.

Rawat, Ramnarayan, and K. Satyanarayana, eds. *Dalit Studies.* Durham, N.C.: Duke University Press, 2016.

Rorty, Richard. *The Linguistic Turn: Recent Essays in Philosophical Method.* Chicago: University of Chicago Press, 1967.

Sahlins, Marshall. *Stone Age Economics.* New York: de Gruyter, 1972.

Said, Edward. *Orientalism.* New York: Pantheon Books, 1978.

Saussure, Ferdinand. *Course in General Linguistics.* Translated by Wade Baskin. New York: Philosophical Library, 1959.

Schleifer, Ronald. *A. J. Greimas and the Nature of Meaning: Linguistics, Semiotics, and Discourse Theory.* London: Croom Helm, 1987.

Shabaraswami, *Shabara-bhashya,* vols. 1–3. Translated by Ganganath Jha. Baroda: University of Baroda Press, 1973. https://archive.org/details /ShabaraBhasyaTrByGanganathJha.

Shklovsky, Viktor. "Art as Technique." In *Russian Formalist Criticism: Four Essays.* Translated by Lee T. Lemon and Marion J. Reis. Lincoln: University of Nebraska Press, 1965.

Simondon, Gilbert. "Technical Mentality." Translated by Arne De Boever. *Parrhesia,* no. 7 (2009): 17–27.

Sonkamble, P. E. *Athvaninche Pakshi.* Aurangabad: Chetna Prakashan, 1979. [Birds of Memory, an autobiography.]

Spivak, Gayatri Chakravorty. *An Aesthetic Education in the Era of Globalization.* Cambridge, Mass.: Harvard University Press, 2012.

———. "Can the Subaltern Speak?" In *Marxism and the Interpretation of Culture,* edited by Cary Nelson and Lawrence Grossberg, 217–313. Urbana: University of Illinois Press, 1988.

———. *A Critique of Postcolonial Reason.* Cambridge, Mass.: Harvard University Press, 1999.

———. *The Death of a Discipline.* New York: Columbia University Press, 2003.

———. *Other Asias*. Malden, Mass.: Blackwell, 2008.

———. *The Postcolonial Critic: Interviews, Strategies, Dialogues*. Edited by Sarah Harasym. New York: Routledge, 1990.

Stanzel, F. K. *A Theory of Narrative*. Translated by Charlotte Goedsche. Preface by Paul Hernadi. Cambridge: Cambridge Paperback Library, 1986.

Stein, Edith. *On the Problem of Empathy*. Translated by Waltraut Stein. Foreword by Erwin W. Straus. The Hague: Martinus Nijhoff, 1964.

Stiegler, Bernard. *Technics and Time*. Vol. 1: *The Fault of Epimetheus*. Translated by Richard Beardsworth and George Collins. Stanford: Stanford University Press, 1998.

———. *Technics and Time*. Vol. 2: *Disorientation*. Translated by S. Barker. Stanford: Stanford University Press, 2008.

Stock, Brian. *The Implications of Literacy: Written Language and Models of Interpretation in the Eleventh and Twelfth Centuries*. Princeton, N.J.: Princeton University Press, 1983.

Sweetman, Will. "The Prehistory of Orientalism: Colonialism and the Textual Basis for Bartholomaus Ziegenbalg's Account of Hinduism." *New Zealand Journal of Asian Studies* 6, no. 2 (2004): 12–38, http://www.otago.ac.nz/religiousstudies/staff/articles/prehistory.pdf.

Tarde, Gabriel. *The Laws of Imitation*. Translated by Elsie Clews Parsons. New York: Henry Holt, 1903. https://ia600303.us.archive.org/14/items/lawsofimitation00tard/lawsofimitation00tard.pdf.

Tolkien, J. R. R. *The Lord of the Rings*. London: George Allen and Unwin, 1968.

Toman, Jindřich. *The Magic of a Common Language: Jakobson, Mathesius, Trubetzkoy, and the Prague Linguistic Circle*. Cambridge, Mass.: MIT Press, 1995.

Trần, Đức Thảo. *Investigations into the Origin of Language and Consciousness*. Translated by Daniel J Herman and Robert L Armstrong. Boston: D. Reidel, 1984.

———. *Phenomenology and Dialectical Materialism*. Edited by Robert S. Cohen. Translated by Daniel J. Herman and Donald V. Morano. Boston: D. Reidel, 1986.

Wakankar, Milind. *Subalternity and Religion: The Prehistory of Dalit Empowerment in South Asia*. London: Routledge, 2010.

Wilberforce, Reginald Garton. *An Unrecorded Chapter of the Indian Mutiny. Being the Personal Reminiscences of R. G. Wilberforce, Late 52nd Light Infantry: Compiled from a Diary and Letters Written on the Spot. With Illustrations*. London: J. Murray, 1894.

Wittgenstein, Ludwig. *Philosophical Investigations*. Translated by G. E. M. Anscombe. Oxford: Basil Blackwell, 1958.

———. *Tractatus Logico-Philosophicus*. Translated by D. F. Pears and Brian F. McGuiness. London: Routledge, 1974.

Xinru, Liu. *Silk and Religion: An Exploration of Material Life and the Thought of People, AD 600–1200*. New York: Oxford University Press, 1996.

Yates, Frances A. *The Art of Memory*. Chicago: University of Chicago Press, 1966.

———. *Giordano Bruno and the Hermetic Tradition*. Chicago: University of Chicago Press, 1964.

———. *The Occult Philosophy in the Elizabethan Age*. London: Routledge and Kegan Paul, 1979.

Žižek, Slavoj. *The Sublime Object of Ideology*. London: Verso, 1989.

INDEX

Acharya, Balkrishna Bapu, 106
Al-Biruni, Muhammad, 144
Alavi, Seema, 89, 107
Althusser, Louis, 134
Ambedkar, B. R., 95, 100, 135–37, 143,
    147–48, 153, 157–59, 176–77, 179, 182, 185,
    188, 192, 194, 197
Annihilation of Caste, 136
Aristotle, 27, 40, 159, 162, 190
Arya Brahma Samaj, 106
Austin, J. L., 154–55

Bagul, Baburao, 7, 142–43, 145–46, 180
Bakhtin, Mikhail, 32, 34, 68–69
Bapat, Ram, 104
Barthes, Roland, 140, 142–43, 168
Bateson, Gregory, 194
Being and Time, 151
Benjamin, Walter, 25, 34, 88, 92, 140–41,
    167
Bentham, Jeremy, 144
Bergson, Henri-Louis, 34
bhakti movement, 71, 130–31
Bhatji, Godse, 49, 78–79, 83–85
Bhatta, Kumaril, 66
Biafra, 185
Blake, William, 62
Bombay, 105–7
Bombay Bible Women's Association, 106
Bombay Branch of Fire Insurance, 106
Bombay National Muhammadan
    Association, 106
Bordieu, Pierre, 24, 118, 174
Bose, H. C., 154
Bosnia, 185
Boswell, John, 46, 74
Brahman Club, 106

Brahman-kanya, 143
Brahmo Samaj, 134
Buddhism, 22, 132–33, 136, 150, 157

Calcutta, 106
Calvin, John, 86
Canara Club, 106
cannibalism, 54
Capital, 50, 167–68
capitalism, 25, 68, 94, 100, 103, 112, 163
Carnap, Rudolf, 155
Centre of Indian Trade Unions (CITU),
    50
Ceremony, 163
Chamber of Commerce, 106
Chamber of Commerce Daily Trade Returns,
    106
Charvaka tradition, 50, 97, 157
Chiplunkar, Vishnushastri, 144
Christ, Jesus, 54
Christianity, 22, 24, 53–54, 63–64, 86, 90,
    97, 133
Comte, Auguste, 174
Cooking the World, 84
Coriolanus, 95
Corporate Social Responsibility, 5
Cotton Traders' Association, 106
Course in General Linguistics, A, 142
*Cracked Mirror: An Indian Debate on
    Experience and Theory, The*, 7
Cultivator's Whip-Cord, The, 127–28, 130–31,
    149, 157
Cusanus, Nicolaus, 1

Daily Commercial Sales Report, 106
Daily Merchants' Companion, 106
Dalit Panther period, 138

dalit philosophy, 52, 112
Dalit Studies, 7
dalit studies, 2, 5–6
Darwin, Charles, 132–33
Dawkins, Richard, 46
Derrida, Jacques, 34, 120, 140, 150, 159, 191, 193
Descartes, René, 162
Dialogic Imagination, The, 32
Dickinson, Emily, 171
Dispossessed, The, 188
Douglas, Mary, 25
Dubois, Abbe, 144
Dumont, Louis, 25

East India Company, 89, 145
Eisenstein, Elizabeth, 140
Eliot, T. S., 63

Faulds, Henry, 154
feminists, 69, 75, 91, 112
Fevre Leuba, 106
Fordism, 113; Post-Fordism, 113
Foucault, Michel, 30, 137, 155, 179; Foucaultian, 152
Frege, Gottlob, 154, 160
Freud, Sigmund, 34, 41
From Sepoy to Subedar, 145
Frye, Northrop, 69
Fukushima, 189

Gadamer, Hans-Georg, 42, 123
Galton, Francis, 154
Gandhi, M. K., 136
gazing, 112
gender, 8, 65, 71, 83
Ghurye, G. S., 149
Gramsci, Antonio, 202
Greaves, 106
Greimas, A. J., 101, 178
Grice, H. P., 155
Guru, Gopal, 7

HPA axis, 54
Haque, Azizul, 154
Hegel, G. W. F., 173
Heidegger, Martin, 2, 27, 34, 42, 114, 151, 154, 177, 195
Henry, Edward, 154–55
Herschel, William, 154
Herzegovina, 185
Hindu Undivided Family, 116
Hinduism, 64, 83, 87, 131–32, 134, 136, 157
Hitchcock, Alfred, 82
Hitler, Adolf, 185

Hokkaido island, 189
Husserl, Edmund, 2, 5, 8, 13, 31, 34, 154, 162; phenomenology, 5, 8, 13, 31

Identity and Difference, 154
Illaiah, Kancha, 52
Implications of Literacy, The, 122
India: British government, 145; Constitution, 136, 153, 188
Islam, 22

Jaimini, 66
Jains, 132
Jakobson, Roman, 142–43
Japan: Ainu, 189
Johnson, Samuel, 144

Kant, Immanuel, 8, 145, 162, 176
Kartsevsky, Sergei, 142
Karve, Iravati, 149
Ketkar, S. V., 143
Kharat, Shankararao, 142
Khmer Rouge, 185
Kindness of Strangers, The, 46
Kripke, Saul, 155

Lacan, Jacques, 33–34, 41, 149
language, 4, 6, 21, 27–28, 37, 41, 54, 62–63, 66, 74, 109, 126, 131, 136, 138, 169, 174, 178, 186–87
Laws of Imitation, 141, 151
Le Guin, Ursula, 188
Lévi-Strauss, Claude, 50, 124, 159, 174
Liu, Xinri, 85
Lord of the Rings, The, 120
Loss of Synaesthesia, 29
Lund and Blokeley, 106
Luther, Martin, 85–86

MacNeice, Louis, 202
Magic of a Common Language, The, 142
Mahabharata, 122, 139
Mahad *satyagraha*, 90
Mahar Regiment, 89–90
Maharashtra, 77, 86–87, 89, 107, 130, 137–38, 144, 168
Malamoud, Charles, 84
Malinowski, Bronislaw, 174
Mandal Commission, 158
Manusmruti, 105, 122
Marathi, 8, 50, 63, 71, 130–31, 136, 138, 142–44, 170, 190
Marx, Karl, 50, 162, 165, 167–68
Mayavati, 159
Maza Pravas, 49, 78

Meinong, Alexius, 154
Merleau-Ponty, Maurice, 13, 31
Mill, J. S., 144
Mill Owners' Association, 106
mimamsa-sutras, 66
Mitakhshara, 86
modernity, 175
Mumbaicha Vruttant, 106

Nagorno-Karabakh, 185
Native Merchants' Association, 106
Nazism, 185
Nirnayasindhu, 86

Oedipus myth, 124
Of Grammatology, 193
On the Problem of Empathy, 39
Ong, Walter, 124
Orality and Literacy, 124
Order Out of Chaos, 141
Orientalists, 127, 144
Oshima, Nagisa, 163

Paine, Tom, 86, 144
Pandey, Sitaram, 145
*Phenomenology and Dialectical
    Materialism*, 97
Phenomenology of Perception, 31
Phillips, 106
Phule, Jyotirao, 85–86, 90, 124, 127–32,
    134–36, 138, 144–45, 147–49, 153, 157–58,
    176–77, 179, 182, 188
Platonic philosophy, 55
Poetics, The, 27, 190
poststructuralism, 8
Pot, Pol, 185
power, 99–100
Prague Linguistic Circle, 142
Prigogine, Ilya, 141
print, 141–42
Psycho, 82
psychopathology, 177
Pure Forms, 55
Purity and Danger, 25

Ramayana, 122, 139
Rawat, Ramnarayan, 7
Reason in the Age of Science, 123
Renaissance, 27
Richardson and Cruddas, 106
Rippon Club, 106
rituals, 55, 86, 96
Rohingya genocide, 185
Romantic lyric poetry, 19

Romanticism, 62, 163
Russel, Bertrand, 155
Rwanda, 185, 201

Sahlins, Marshall, 164
Sapporo, 189
Sarukkai, Sundar, 7
Satyanarayana, K., 7
Saussure, Ferdinand, 142–43, 178
Searle, John, 155
Selfish Gene, The, 46
Sepoys and the Company, The, 89
Shabarabhashya, 66
Shakespeare, William, 95
Shankaracharya, 131
Shetkaryacha Asud, 85
Shingane, Moro Vinayak, 106
shramana movement, 71
Silk and Religion, 85
Simondon, Gilbert, 162, 165–66
sociability, 170–74, 180, 183, 198
Society Must Be Defended, 179
Sonkamble, P. E., 142
Sood, 145
Stanzel, Franz, 38
Stein, Edith, 39
Stengers, Isabelle, 141
Stock, Brian, 122
Stone Age Economics, 164

Tansen, 133
Tarde, Gabriel, 141, 151, 175
Taylorism, 113
Thảo, Trần Đức, 28, 97
Theory of Narrative, 38
Toman, Jindřich, 142
Traders' Association, 106
Trojan, Stefanie, 173–74
Trubetzkoy, Nikolai, 142

Underwriters' Association, 106

Veblen, Thorstein, 24
Vedas, 119, 134, 141
Vedic tradition, 50

Who Were the Shudras?, 95
Why I Am Not a Hindu, 52
Wittgenstein, Ludwig, 27, 155

Yemen, 189

Ziegenbalg, Bartholomaus, 144
Žižek, Slavoj, 33

Roberto Esposito, *Terms of the Political: Community, Immunity, Biopolitics.* Translated by Rhiannon Noel Welch. Introduction by Vanessa Lemm.

Maurizio Ferraris, *Documentality: Why It Is Necessary to Leave Traces.* Translated by Richard Davies.

Dimitris Vardoulakis, *Sovereignty and Its Other: Toward the Dejustification of Violence.*

Anne Emmanuelle Berger, *The Queer Turn in Feminism: Identities, Sexualities, and the Theater of Gender.* Translated by Catherine Porter.

James D. Lilley, *Common Things: Romance and the Aesthetics of Belonging in Atlantic Modernity.*

Jean-Luc Nancy, *Identity: Fragments, Frankness.* Translated by François Raffoul.

Miguel Vatter, *Between Form and Event: Machiavelli's Theory of Political Freedom.*

Miguel Vatter, *The Republic of the Living: Biopolitics and the Critique of Civil Society.*

Maurizio Ferraris, *Where Are You? An Ontology of the Cell Phone.* Translated by Sarah De Sanctis.

Irving Goh, *The Reject: Community, Politics, and Religion after the Subject.*

Kevin Attell, *Giorgio Agamben: Beyond the Threshold of Deconstruction.*

J. Hillis Miller, *Communities in Fiction.*

Remo Bodei, *The Life of Things, the Love of Things*. Translated by Murtha Baca.

Gabriela Basterra, *The Subject of Freedom: Kant, Levinas*.

Roberto Esposito, *Categories of the Impolitical*. Translated by Connal Parsley.

Roberto Esposito, *Two: The Machine of Political Theology and the Place of Thought*. Translated by Zakiya Hanafi.

Akiba Lerner, *Redemptive Hope: From the Age of Enlightenment to the Age of Obama*.

Adriana Cavarero and Angelo Scola, *Thou Shalt Not Kill: A Political and Theological Dialogue*. Translated by Margaret Adams Groesbeck and Adam Sitze.

Massimo Cacciari, *Europe and Empire: On the Political Forms of Globalization*. Edited by Alessandro Carrera, Translated by Massimo Verdicchio.

Emanuele Coccia, *Sensible Life: A Micro-ontology of the Image*. Translated by Scott Stuart, Introduction by Kevin Attell.

Timothy C. Campbell, *The Techne of Giving: Cinema and the Generous Forms of Life*.

Étienne Balibar, *Citizen Subject: Foundations for Philosophical Anthropology*. Translated by Steven Miller, Foreword by Emily Apter.

Ashon T. Crawley, *Blackpentecostal Breath: The Aesthetics of Possibility*.

Terrion L. Williamson, *Scandalize My Name: Black Feminist Practice and the Making of Black Social Life*.

Jean-Luc Nancy, *The Disavowed Community*. Translated by Philip Armstrong.

Roberto Esposito, *The Origin of the Political: Hannah Arendt or Simone Weil?* Translated by Vincenzo Binetti and Gareth Williams.

Dimitris Vardoulakis, *Stasis before the State: Nine Theses on Agonistic Democracy*.

Nicholas Heron, *Liturgical Power: Between Economic and Political Theology.*

Emanuele Coccia, *Goods: Advertising, Urban Space, and the Moral Law of the Image.* Translated by Marissa Gemma.

Aniket Jaaware, *Practicing Caste: On Touching and Not Touching.*

www.ingramcontent.com/pod-product-compliance
Lightning Source LLC
Chambersburg PA
CBHW032129020426
42334CB00016B/1088